EurOtop

Wave Overtopping of Sea Defences and Related Structures: Assessment Manual

August 2007

EA   Environment Agency, UK
ENW Expertise Netwerk Waterkeren, NL
KFKI Kuratorium für Forschung im Küsteningenieurwesen, DE

www.overtopping-manual.com

The EurOtop Team

Authors:
T. Pullen (HR Wallingford, UK)
N. W. H. Allsop (HR Wallingford, UK)
T. Bruce (University Edinburgh, UK)
A. Kortenhaus (Leichtweiss Institut, DE)
H. Schüttrumpf (Bundesanstalt für Wasserbau, DE)
J. W. van der Meer (Infram, NL)

Steering group:
C. Mitchel (Environment Agency/DEFRA, UK)
M. Owen (Environment Agency/DEFRA, UK)
D. Thomas (Independent Consultant; Faber Maunsell, UK)
P. van den Berg (Hoogheemraadschap Rijnland, NL – till 2006)
H. van der Sande (Waterschap Zeeuwse Eilanden, NL – from 2006)
M. Klein Breteler (WL | Delft Hydraulics, NL)
D. Schade (Ingenieursbüro Mohn GmbH, DE)

Funding bodies:
This manual was funded in the UK by the Environmental Agency, in Germany by the German Coastal Engineering Research Council (KFKI), and in the Netherlands by Rijkswaterstaat, Netherlands Expertise Network on Flood Protection.

This manual replaces:
EA, 1999. Overtopping of Seawalls. Design and Assessment Manual, HR, Wallingford Ltd, R&D Technical Report W178. Author: P. Besley.

TAW, 2002. Technical Report Wave Run-up and Wave Overtopping at Dikes. TAW, Technical Advisory Committee on Flood Defences. Author: J. W. van der Meer.

EAK, 2002. Ansätze für die Bemessung von Küstenschutzwerken. Chapter 4 in Die Küste, Archive for Research and Technology on the North Sea and Baltic Coast. Empfehlungen für Küstenschutzwerke.

Die Küste

ARCHIV FÜR FORSCHUNG UND TECHNIK
AN DER NORD- UND OSTSEE

ARCHIVE FOR RESEARCH AND TECHNOLOGY
ON THE NORTH SEA AND BALTIC COAST

Heft 73 · Jahr 2007

Herausgeber: Kuratorium für Forschung im Küsteningenieurwesen

**EurOtop
Wave Overtopping of Sea Defences
and Related Structures:
Assessment Manual**

Kommissionsverlag:
Boyens Medien GmbH & Co. KG, Heide i. Holstein
Druck: Boyens Offset

ISSN 0452-7739
ISBN 978-3-8042-1064-6

Anschriften der Verfasser dieses Heftes:

The EurOtop Team
T. Pullen (HR Wallingford, UK); N.W.H. Allsop (HR Wallingford, UK); T. Bruce (University Edinburgh, UK); A. Kortenhaus (Leichtweiss Institut, DE); H. Schüttrumpf (Bundesanstalt für Wasserbau, DE); J. W. van der Meer (Infram, NL).

Die Verfasser sind für den Inhalt der Aufsätze allein verantwortlich. Nachdruck aus dem Inhalt nur mit Genehmigung des Herausgebers gestattet: Kuratorium für Forschung im Küsteningenieurwesen, Geschäftsstelle, Wedeler Landstraße 157, 22559 Hamburg.
Vorsitzender des Kuratoriums: MR BERND PROBST, Mercatorstraße 3, 24106 Kiel
Geschäftsführer: Dr.-Ing. RAINER LEHFELDT, Wedeler Landstraße 157, 22559 Hamburg
Schriftleitung „Die Küste": Dr.-Ing. VOLKER BARTHEL, Birkenweg 6a, 27607 Langen

Preface

Why is this Manual needed?

This Overtopping Manual gives guidance on analysis and/or prediction of wave overtopping for flood defences attacked by wave action. It is primarily, but not exclusively, intended to assist government, agencies, businesses and specialist advisors & consultants concerned with reducing flood risk. Methods and guidance described in the manual may also be helpful to designers or operators of breakwaters, reclamations, or inland lakes or reservoirs.

Developments close to the shoreline (coastal, estuarial or lakefront) may be exposed to significant flood risk yet are often highly valued. Flood risks are anticipated to increase in the future driven by projected increases of sea levels, more intense rainfall, and stronger wind speeds. Levels of flood protection for housing, businesses or infrastructure are inherently variable. In the Netherlands, where two-thirds of the country is below storm surge level, large rural areas may presently (2007) be defended to a return period of 1:10,000 years, with less densely populated areas protected to 1:4,000 years. In the UK, where low-lying areas are much smaller, new residential developments are required to be defended to 1:200 year return.

Understanding future changes in flood risk from waves overtopping seawalls or other structures is a key requirement for effective management of coastal defences. Occurrences of economic damage or loss of life due to the hazardous nature of wave overtopping are more likely, and coastal managers and users are more aware of health and safety risks. Seawalls range from simple earth banks through to vertical concrete walls and more complex composite structures. Each of these require different methods to assess overtopping.

Reduction of overtopping risk is therefore a key requirement for the design, management and adaptation of coastal structures, particularly as existing coastal infrastructure is assessed for future conditions. There are also needs to warn or safeguard individuals potentially to overtopping waves on coastal defences or seaside promenades, particularly as recent deaths in the UK suggest significant lack of awareness of potential dangers.

Guidance on wave run-up and overtopping have been provided by previous manuals in UK, Netherlands and Germany including the EA Overtopping Manual edited by Besley (1999); the TAW Technical Report on Wave run up and wave overtopping at dikes by van der Meer (2002); and the German Die Küste EAK (2002). Significant new information has now been obtained from the EC CLASH project collecting data from several nations, and further advances from national research projects. This Manual takes account of this new information and advances in current practice. In so doing, this manual will extend and/or revise advice on wave overtopping predictions given in the CIRIA/CUR Rock Manual, the Revetment Manual by McConnell (1998), British Standard BS6349, the US Coastal Engineering Manual, and ISO TC98.

The Manual and Calculation Tool

The Overtopping Manual incorporates new techniques to predict wave overtopping at seawalls, flood embankments, breakwaters and other shoreline structures. The manual includes case studies and example calculations. The manual has been intended to assist coastal engineers analyse overtopping performance of most types of sea defence found around Europe. The methods in the manual can be used for current performance assessments and for longer-term design calculations. The manual defines types of structure, provides definitions for parameters, and gives guidance on how results should be interpreted. A chapter on hazards gives guidance on tolerable discharges and overtopping processes. Further discussion identifies the different methods available for assessing overtopping, such as empirical, physical and numerical techniques.

In parallel with this manual, an online Calculation Tool has been developed to assist the user through a series of steps to establish overtopping predictions for: embankments and dikes; rubble mound structures; and vertical structures. By selecting an indicative structure type and key structural features, and by adding the dimensions of the geometric and hydraulic parameters, the mean overtopping discharge will be calculated. Where possible additional results for overtopping volumes, flow velocities and depths, and other pertinent results will be given.

Intended use

The manual has been intended to assist engineers who are already aware of the general principles and methods of coastal engineering. The manual uses methods and data from research studies around Europe and overseas so readers are expected to be familiar with wave and response parameters and the use of empirical equations for prediction. Users may be concerned with existing defences, or considering possible rehabilitation or new-build.

This manual is not, however, intended to cover many other aspects of the analysis, design, construction or management of sea defences for which other manuals and methods already exist, see for example the CIRIA/CUR/CETMEF Rock Manual (2007), the Beach Management Manual by BRAMPTON et al. (2002) and TAW guidelines in the Netherlands on design of sea, river and lake dikes.

What next?

It is clear that increased attention to flood risk reduction, and to wave overtopping in particular, have increased interest and research in this area. This Manual is, therefore, not expected to be the 'last word' on the subject, indeed even whilst preparing this version, it was expected that there will be later revisions. At the time of writing this preface (August 2007), we anticipate that there may be sufficient new research results available to justify a further small revision of the Manual in the summer or autumn of 2008.

The Authors and Steering Committee
August 2007

Content

The Europe Team	IV
Preface	V
1. Introduction	1
1.1 Background	1
1.1.1 Previous and related manuals	1
1.1.2 Sources of material and contributing projects	1
1.2 Use of this manual	1
1.3 Principal types of structures	2
1.4 Definitions of key parameters and principal responses	3
1.4.1 Wave height	4
1.4.2 Wave period	4
1.4.3 Wave steepness and Breaker parameter	5
1.4.4 Parameter h_*	7
1.4.5 Toe of structure	7
1.4.6 Foreshore	7
1.4.7 Slope	8
1.4.8 Berm	8
1.4.9 Crest freeboard and armour freeboard and width	8
1.4.10 Permeability, porosity and roughness	10
1.4.11 Wave run-up height	11
1.4.12 Wave overtopping discharge	12
1.4.13 Wave overtopping volumes	13
1.5 Probability levels and uncertainties	13
1.5.1 Definitions	13
1.5.2 Background	14
1.5.3 Parameter uncertainty	16
1.5.4 Model uncertainty	16
1.5.5 Methodology and output	17
2. Water levels and wave conditions	18
2.1 Introduction	18
2.2 Water levels, tides, surges and sea level changes	18
2.2.1 Mean sea level	18
2.2.2 Astronomical tide	19
2.2.3 Surges related to extreme weather conditions	19
2.2.4 High river discharges	20
2.2.5 Effect on crest levels	21
2.3 Wave conditions	22
2.4 Wave conditions at depth-limited situations	23
2.5 Currents	26
2.6 Application of design conditions	27
2.7 Uncertainties in inputs	28
3. Tolerable discharges	29
3.1 Introduction	29
3.1.1 Wave overtopping processes and hazards	29
3.1.2 Types of overtopping	30
3.1.3 Return periods	31
3.2 Tolerable mean discharges	32
3.3 Tolerable maximum volumes and velocities	36
3.3.1 Overtopping volumes	36
3.3.2 Overtopping velocities	37
3.3.3 Overtopping loads and overtopping simulator	37
3.4 Effects of debris and sediment in overtopping flows	39

4. Prediction of overtopping 40
- 4.1 Introduction 40
- 4.2 Empirical models, including comparison of structures 40
  - 4.2.1 Mean overtopping discharge 40
  - 4.2.2 Overtopping volumes and Vmax 44
  - 4.2.3 Wave transmission by wave overtopping 46
- 4.3 PC-OVERTOPPING 51
- 4.4 Neural network tools 55
- 4.5 Use of CLASH database 60
- 4.6 Outline of numerical model types 62
  - 4.6.1 Navier-Stokes models 63
  - 4.6.2 Nonlinear shallow water equation models 63
- 4.7 Physical modelling 64
- 4.8 Model and Scale effects 65
  - 4.8.1 Scale effects 65
  - 4.8.2 Model and measurement effects 65
  - 4.8.3 Methodology 66
- 4.9 Uncertainties in predictions 67
  - 4.9.1 Empirical Models 67
  - 4.9.2 Neural Network 68
  - 4.9.3 CLASH database 68
- 4.10 Guidance on use of methods 68

5. Coastal dikes and embankment seawalls 70
- 5.1 Introduction 70
- 5.2 Wave run-up 71
  - 5.2.1 History of the 2% value for wave run-up 77
- 5.3 Wave overtopping discharges 77
  - 5.3.1 Simple slopes 77
  - 5.3.2 Effect of roughness 85
  - 5.3.3 Effect of oblique waves 90
  - 5.3.4 Composite slopes and berms 93
  - 5.3.5 Effect of wave walls 97
- 5.4 Overtopping volumes 99
- 5.5 Overtopping flow velocities and overtopping flow depth 100
  - 5.5.1 Seaward Slope 102
  - 5.5.2 Dike Crest 103
  - 5.5.3 Landward Slope 107
- 5.6 Scale effects for dikes 109
- 5.7 Uncertainties 109

6. Armoured rubble slopesa and mounds 111
- 6.1 Introduction 111
- 6.2 Wave run-up and run-down levels, number of overtopping waves 112
- 6.3 Overtopping discharges 117
  - 6.3.1 Simple armoured slopes 117
  - 6.3.2 Effect of armoured crest berm 120
  - 6.3.3 Effect of oblique waves 120
  - 6.3.4 Composite slopes and berms, including berm breakwaters 121
  - 6.3.5 Effect of wave walls 124
  - 6.3.6 Scale and model effect corrections 125
- 6.4 Overtopping volumes per wave 126
- 6.5 Overtopping velocities and spatial distribution 127
- 6.6 Overtopping of shingle beaches 129
- 6.7 Uncertainties 129

7. Vertical and steep seawalls 130
- 7.1 Introduction 130
- 7.2 Wave processes at walls 132

      7.2.1 Overview 132
      7.2.2 Overtopping regime discrimination – plain vertical walls 134
      7.2.3 Overtopping regime discrimination – composite vertical walls 135
  7.3 Mean overtopping discharges for vertical and battered walls 136
      7.3.1 Plain vertical walls 136
      7.3.2 Battered walls 140
      7.3.3 Composite vertical walls 141
      7.3.4 Effect of oblique waves 142
      7.3.5 Effect of bullnose and recurve walls 145
      7.3.6 Effect of wind 148
      7.3.7 Scale and model effect corrections 149
  7.4 Overtopping volumes 151
      7.4.1 Introduction 151
      7.4.2 Overtopping volumes at plain vertical walls 151
      7.4.3 Overtopping volumes at composite (bermed) structures 153
      7.4.4 Overtopping volumes at plain vertical walls under oblique wave attack . . . 153
      7.4.5 Scale effects for individual overtopping volumes 154
  7.5 Overtopping velocities, distributions and down-fall pressures 154
      7.5.1 Introduction to post-overtopping processes 154
      7.5.2 Overtopping throw speeds 154
      7.5.3 Spatial extent of overtopped discharge 155
      7.5.4 Pressures resulting from downfalling water mass 156
  7.6 Uncertainties 157

Glossary 158

Notation 160

References 164

A Structure of the EurOtop calculation tool 173

Figures

Figure 1.1:	Type of breaking on a slope	5
Figure 1.2:	Spilling waves on a beach; $\xi_{m-1,0} < 0.2$.	6
Figure 1.3:	Plunging waves; $\xi_{m-1,0} < 2.0$.	6
Figure 1.4:	Crest freeboard different from armour freeboard	9
Figure 1.5:	Crest freeboard ignores a permeable layer if no crest element is present.	9
Figure 1.6:	Crest configuration for a vertical wall	10
Figure 1.7:	Example of wave overtopping measurements, showing the random behaviour	12
Figure 1.8:	Sources of uncertainties	15
Figure 1.9:	Gaussian distribution function and variation of parameters	15
Figure 2.1:	Measurements of maximum water levels for more than 100 years and extrapolation to extreme return periods	20
Figure 2.2:	Important aspects during calculation or assessment of dike height.	21
Figure 2.3:	Wave measurements and numerical simulations in the North Sea (1964–1993), leading to an extreme distribution	22
Figure 2.4:	Depth-limited significant wave heights for uniform foreshore slopes	24
Figure 2.5:	Computed composite Weibull distribution. $H_{m0} = 3.9$ m; foreshore slope 1:40 and water depth h = 7 m.	26
Figure 2.6:	Encounter probability	27
Figure 3.1:	Overtopping on embankment and promenade seawalls.	31
Figure 3.2:	Wave overtopping test on bare clay; result after 6 hours with 10 l/s per m width	36
Figure 3.3:	Example wave forces on a secondary wall	37
Figure 3.4:	Principle of the wave overtopping simulator.	38
Figure 3.5:	The wave overtopping simulator discharging a large overtopping volume on the inner slope of a dike	39
Figure 4.1:	Comparison of wave overtopping formulae for various kind of structures	43
Figure 4.2:	Comparison of wave overtopping as function of slope angle.	43
Figure 4.3:	Various distributions on a Rayleigh scale graph. A straight line (b = 2) is a Rayleigh distribution.	44
Figure 4.4:	Relationship between mean discharge and maximum overtopping volume in one wave for smooth, rubble mound and vertical structures for wave heights of 1 m and 2.5 m.	46
Figure 4.5:	Wave transmission for a gentle smooth structure of 1:4 and for different wave steepness	47
Figure 4.6:	Wave overtopping for a gentle smooth structure of 1:4 and for different wave steepness	48
Figure 4.7:	Wave transmission versus wave overtopping for a smooth 1:4 slope and a wave height of $H_{m0} = 3$ m.	48
Figure 4.8:	Wave transmission versus wave overtopping discharge for a rubble mound structure, $\cot\alpha = 1.5$; 6-10 ton rock, B = 4.5 m and $H_{m0} = 3$ m	49
Figure 4.9:	Comparison of wave overtopping and transmission for a vertical, rubble mound and smooth structure	50
Figure 4.10:	Wave overtopping and transmission at breakwater IJmuiden, the Netherlands.	51
Figure 4.11:	Example cross-section of a dike.	52
Figure 4.12:	Input of geometry by x-y coordinates and choice of top material	52
Figure 4.13:	Input file	53
Figure 4.14:	Output of pc-overtopping.	53
Figure 4.15:	Check on 2%-runup level.	54
Figure 4.16:	Check on mean overtopping discharge.	54
Figure 4.17:	Configuration of the neural network for wave overtopping	56
Figure 4.18:	Overall view of possible structure configurations for the neural network.	57
Figure 4.19:	Example cross-section with parameters for application of neural network	58
Figure 4.20:	Results of a trend calculation	59
Figure 4.21:	Overtopping for large wave return walls; first selection.	61

Figure 4.22:	Overtopping for large wave return walls; second selection with more criteria	61
Figure 4.23:	Overtopping for a wave return wall with $s_o = 0.04$, seaward angle of 45°, a width of 2 m and a crest height of $R_c = 3$ m. For $H_{m0\,toe} = 3$ m the overtopping can be estimated from $R_c/H_{m0\,toe} = 1$.	62
Figure 5.1:	Wave run-up and wave overtopping for coastal dikes and embankment seawalls: definition sketch. See Section 1.4 for definitions.	70
Figure 5.2:	Main calculation procedure for coastal dikes and embankment seawalls. . .	71
Figure 5.3:	Definition of the wave run-up height $R_{u2\%}$ on a smooth impermeable slope.	72
Figure 5.4:	Relative Wave run-up height $R_{u2\%}/H_{m0}$ as a function of the breaker parameter $\xi_{m-1,0}$, for smooth straight slopes	73
Figure 5.5:	Relative Wave run-up height $R_{u2\%}/H_{m0}$ as a function of the wave steepness for smooth straight slopes	73
Figure 5.6:	Wave run-up for smooth and straight slopes.	75
Figure 5.7:	Wave run-up for deterministic and probabilistic design.	76
Figure 5.8:	Wave overtopping as a function of the wave steepness H_{m0}/L_0 and the slope	78
Figure 5.9:	Wave overtopping data for breaking waves and overtopping Equation 5.8 with 5 % under and upper exceedance limits.	79
Figure 5.10:	Wave overtopping data for non-breaking waves and overtopping Equation 5.9 with 5 % under and upper exceedance limits	80
Figure 5.11:	Wave overtopping for breaking waves – Comparison of formulae for design and safety assessment and probabilistic calculations.	81
Figure 5.12:	Wave overtopping for non-breaking waves – Comparison of formulae for design and safety assessment and probabilistic calculations.	81
Figure 5.13:	Dimensionless overtopping discharge for zero freeboard (SCHÜTTRUMPF, 2001)	84
Figure 5.14:	Wave overtopping and overflow for positive, zero and negative freeboard .	84
Figure 5.15:	Dike covered by grass (photo: Schüttrumpf).	85
Figure 5.16:	Dike covered by asphalt (photo: Schüttrumpf)	86
Figure 5.17:	Dike covered by natural bloc revetment (photo: Schüttrumpf).	86
Figure 5.18:	Influence factor for grass surface	87
Figure 5.19:	Example for roughness elements (photo: Schüttrumpf).	88
Figure 5.20:	Dimensions of roughness elements	89
Figure 5.21:	Performance of roughness elements showing the degree of turbulence . . .	90
Figure 5.22:	Definition of angle of wave attack β	91
Figure 5.23:	Short crested waves resulting in wave run-up and wave overtopping (photo: Zitscher)	92
Figure 5.24:	Influence factor γ_β for oblique wave attack and short crested waves, measured data are for wave run-up	93
Figure 5.25:	Determination of the average slope (1st estimate)	94
Figure 5.26:	Determination of the average slope (2nd estimate)	94
Figure 5.27:	Determination of the characteristic berm length L_{Berm}	95
Figure 5.28:	Typical berms (photo: Schüttrumpf)	95
Figure 5.29:	Influence of the berm depth on factor r_{dh}	97
Figure 5.30:	Sea dike with vertical crest wall (photo: Hofstede)	97
Figure 5.31:	Influence of a wave wall on wave overtopping (photo: Schüttrumpf)	98
Figure 5.32:	Example probability distribution for wave overtopping volumes per wave .	101
Figure 5.33:	Wave overtopping on the landward side of a seadike (photo: Zitscher) . . .	101
Figure 5.34:	Definition sketch for layer thickness and wave run-up velocities on the seaward slope.	102
Figure 5.35:	Wave run-up velocity and wave run-up flow depth on the seaward slope (example)	104
Figure 5.36:	Sequence showing the transition of overtopping flow on a dike crest (Large Wave Flume, Hannover).	104
Figure 5.37:	Definition sketch for overtopping flow parameters on the dike crest	105
Figure 5.38:	Overtopping flow velocity data compared to the overtopping flow velocity formula	106

Figure 5.39:	Sensitivity analysis for the dike crest (left side: influence of overtopping flow depth on overtopping flow velocity; right side: influence of bottom friction on overtopping flow velocity)	106
Figure 5.40:	Overtopping flow on the landward slope (Large Wave Flume, Hannover) (photo: Schüttrumpf)	107
Figure 5.41:	Definition of overtopping flow parameters on the landward slope	108
Figure 5.42:	Sensitivity Analysis for Overtopping flow velocities and related overtopping flow depths – Influence of the landward slope –	109
Figure 5.43:	Wave overtopping over sea dikes, including results from uncertainty calculations	110
Figure 6.1:	Armoured structures	112
Figure 6.2:	Relative run-up on straight rock slopes with permeable and impermeable core, compared to smooth impermeable slopes	113
Figure 6.3:	Run-up level and location for overtopping differ	115
Figure 6.4:	Percentage of overtopping waves for rubble mound breakwaters as a function of relative (armour) crest height and armour size ($R_c \leq A_c$)	116
Figure 6.5:	Relative 2 % run-down on straight rock slopes with impermeable core (imp), permeable core (perm) and homogeneous structure (hom)	117
Figure 6.6:	Mean overtopping discharge for 1:1.5 smooth and rubble mound slopes	119
Figure 6.7	Icelandic Berm breakwater	121
Figure 6.8:	Conventional reshaping berm breakwater	122
Figure 6.9:	Non-reshaping Icelandic berm breakwater with various classes of big rock	122
Figure 6.10:	Proposed adjustment factor applied to data from two field sites (Zeebrugge 1:1.4 rubble mound breakwater, and Ostia 1:4 rubble slope)	126
Figure 6.11:	Definition of y for various cross-sections	127
Figure 6.12:	Definition of x- and y-coordinate for spatial distribution	128
Figure 7.1:	Examples of vertical breakwaters: (left) modern concrete caisson and (right) older structure constructed from concrete blocks	130
Figure 7.2:	Examples of vertical seawalls: (left) modern concrete wall and (right) older stone blockwork wall	130
Figure 7.3:	A non-impulsive (pulsating) wave condition at a vertical wall, resulting in non-impulsive (or "green water") overtopping	133
Figure 7.4:	An impulsive (breaking) wave at a vertical wall, resulting in an impulsive (violent) overtopping condition	133
Figure 7.5:	A broken wave at a vertical wall, resulting in a broken wave overtopping condition	133
Figure 7.6:	Definition sketch for assessment of overtopping at plain vertical walls	134
Figure 7.7:	Definition sketch for assessment of overtopping at composite vertical walls	135
Figure 7.8:	Mean overtopping at a plain vertical wall under non-impulsive conditions (Equations 7.3 and 7.4)	136
Figure 7.9:	Dimensionless overtopping discharge for zero freeboard (SMID, 2001)	137
Figure 7.10:	Mean overtopping at a plain vertical wall under impulsive conditions (Equations 7.6 and 7.7)	138
Figure 7.11:	Mean overtopping discharge for lowest $h* R_c / H_{m0}$ (for broken waves only arriving at wall) with submerged toe ($h_s > 0$). For $0.02 < h* R_c / H_{m0} < 0.03$, overtopping response is ill-defined – lines for both impulsive conditions (extrapolated to lower $h* R_c / H_{m0}$) and broken wave only conditions (extrapolated to higher $h* R_c / H_{m0}$) are shown as dashed lines over this region	139
Figure 7.12:	Mean overtopping discharge with emergent toe ($h_s < 0$)	140
Figure 7.13:	Battered walls: typical cross-section (left), and Admiralty Breakwater, Alderney Channel Islands (right, courtesy G. Müller)	141
Figure 7.14:	Overtopping for a 10:1 and 5:1 battered walls	141
Figure 7.15:	Overtopping for composite vertical walls	143
Figure 7.16:	Overtopping of vertical walls under oblique wave attack	144
Figure 7.17:	An example of a modern, large vertical breakwater with wave return wall (left) and cross-section of an older seawall with recurve (right)	145
Figure 7.18:	A sequence showing the function of a parapet / wave return wall in reducing overtopping by redirecting the uprushing water seaward (back to right)	145

Figure 7.19: Parameter definitions for assessment of overtopping at structures with parapet/wave return wall. 146
Figure 7.20: "Decision chart" summarising methodology for tentative guidance. Note that symbols R_0^*, k_{23}, m and m^* used (only) at intermediate stages of the procedure are defined in the lowest boxes in the figure. Please refer to text for further explanation.. 147
Figure 7.21: Wind adjustment factor f_{wind} plotted over mean overtopping rates q_{ss} 148
Figure 7.22: Large-scale laboratory measurements of mean discharge at 10:1 battered wall under impulsive conditions showing agreement with prediction line based upon small-scale tests (Equation 7.12) 150
Figure 7.23: Results from field measurements of mean discharge at Samphire Hoe, UK, plotted together with Equation 7.13 150
Figure 7.24: Predicted and measured maximum individual overtopping volume – small- and large-scale tests (PEARSON et al., 2002). 152
Figure 7.25: Speed of upward projection of overtopping jet past structure crest plotted with "impulsiveness parameter" $h*$ (after BRUCE et al., 2002) 155
Figure 7.26: Landward distribution of overtopping discharge under impulsive conditions. Curves show proportion of total overtopping discharge which has landed within a particular distance shoreward of seaward crest . 156

Tables

Table 2.1:	Values of dimensionless wave heights for some values of H_{tr}/H_{rms}	25
Table 3.1:	Hazard Type	32
Table 3.2:	Limits for overtopping for pedestrians	33
Table 3.3:	Limits for overtopping for vehicles	34
Table 3.4:	Limits for overtopping for property behind the defence	34
Table 3.5:	Limits for overtopping for damage to the defence crest or rear slope	35
Table 4.1:	Example input file for neural network with first 6 calculations	58
Table 4.2:	Output file of neural network with confidence limits	58
Table 4.3:	Scale effects and critical limits	67
Table 5.1:	Owen's coefficients for simple slopes	82
Table 5.2:	Surface roughness factors for typical elements	88
Table 5.3:	Characteristic values for parameter c_2 (TMA-spectra)	102
Table 5.4:	Characteristic Values for Parameter a_0^* (TMA-spectra)	103
Table 6.1:	Main calculation procedure for armoured rubble slopes and mounds	111
Table 6.2:	Values for roughness factor γ_f for permeable rubble mound structures with slope of 1:1.5. Values in italics are estimated/extrapolated	119
Table 7.1:	Summary of principal calculation procedures for vertical structures	132
Table 7.2:	Summary of prediction formulae for individual overtopping volumes under oblique wave attack. Oblique cases valid for $0.2 < h* R_c / H_{m0} < 0.65$. For $0.07 < h* R_c / H_{m0} < 0.2$, the $\beta = 0°$ formulae should be used for all β	153
Table 7.3:	Probabilistic and deterministic design parameters for vertical and battered walls	157

1. Introduction

1.1 Background

This manual describes methods to predict wave overtopping of sea defence and related coastal or shoreline structures. It recommends approaches for calculating mean overtopping discharges, maximum overtopping volumes and the proportion of waves overtopping a seawall. The manual will help engineers to establish limiting tolerable discharges for design wave conditions, and then use the prediction methods to confirm that these discharges are not exceeded.

1.1.1 Previous and related manuals

This manual is developed from, at least in part, three manuals: the (UK) Environment Agency Manual on Overtopping edited by BESLEY (1999); the (Netherlands) TAW Technical Report on Wave run-up and wave overtopping at dikes, edited by VAN DER MEER (2002); and the German Die Küste EAK (2002) edited by ERCHINGER. The new combined manual is intended to revise, extend and develop the parts of those manuals discussing wave run-up and overtopping.

In so doing, this manual will also extend and/or revise advice on wave overtopping predictions given in the CIRIA / CUR Rock Manual, the Revetment Manual by MCCONNELL (1998), British Standard BS6349, the US Coastal Engineering Manual, and ISO TC98.

1.1.2 Sources of material and contributing projects

Beyond the earlier manuals discussed in section 1.3, new methods and data have been derived from a number of European and national research programmes. The main new contributions to this manual have been derived from OPTICREST; PROVERBS; CLASH & SHADOW, VOWS and Big-VOWS and partly ComCoast. Everything given in this manual is supported by research papers and manuals described in the bibliography.

1.2 Use of this manual

The manual has been intended to assist an engineer analyse the overtopping performance of any type of sea defence or related shoreline structure found around Europe. The manual uses the results of research studies around Europe and further overseas to predict wave overtopping discharges, number of overtopping waves, and the distributions of overtopping volumes. It is envisaged that methods described here may be used for current performance assessments, and for longer-term design calculations. Users may be concerned with existing defences, or considering possible rehabilitation or new-build.

The analysis methods described in this manual are primarily based upon a deterministic approach in which overtopping discharges (or other responses) are calculated for wave and water level conditions representing a given return period. All of the design equations require data on water levels and wave conditions at the toe of the defence structure. The input water level should include a tidal and, if appropriate, a surge component. Surges are usually comprised of components including wind set-up and barometric pressure. Input wave conditions should take account of nearshore wave transformations, including breaking. Methods of calculating depth-limited wave conditions are outlined in Chapter 2.

All of the prediction methods given in this report have intrinsic limitations to their accuracy. For empirical equations derived from physical model data, account should be taken of the inherent scatter. This scatter, or reliability of the equations, has been described where possible or available and often equations for deterministic use are given where some safety has been taken into account. Still it can be concluded that overtopping rates calculated by empirically derived equations, should only be regarded as being within, at best, a factor of 1–3 of the actual overtopping rate. The largest deviations will be found for small overtopping discharges.

As however many practical structures depart (at least in part) from the idealised versions tested in hydraulics laboratories, and it is known that overtopping rates may be very sensitive to small variations in structure geometry, local bathymetry and wave climate, empirical methods based upon model tests conducted on generic structural types, such as vertical walls, armoured slopes etc may lead to large differences in overtopping performance. Methods presented here will not predict overtopping performance with the same degree of accuracy as structure-specific model tests.

This manual is not however intended to cover many other aspects of the analysis, design, construction or management of sea defences for which other manuals and methods already exist, see for example CIRIA/CUR (1991), BSI (1991), SIMM et al. (1996), BRAMPTON et al. (2002) and TAW guidelines in the Netherlands on design of sea, river and lake dikes. The manual has been kept deliberately concise in order to maintain clarity and brevity. For the interested reader a full set of references is given so that the reasoning behind the development of the recommended methods can be followed.

1.3 Principal types of structures

Wave overtopping is of principal concern for structures constructed primarily to defend against flooding: often termed sea defence. Somewhat similar structures may also be used to provide protection against coastal erosion: sometimes termed coast protection. Other structures may be built to protect areas of water for ship navigation or mooring: ports, harbours or marinas; these are often formed as breakwaters or moles. Whilst some of these types of structures may be detached from the shoreline, sometimes termed offshore, nearshore or detached, most of the structures used for sea defence form a part of the shoreline.

This manual is primarily concerned with the three principal types of sea defence structures: sloping sea dikes and embankment seawalls; armoured rubble slopes and mounds; and vertical, battered or steep walls.

Historically, sloping dikes have been the most widely used option for sea defences along the coasts of the Netherlands, Denmark, Germany and many parts of the UK. Dikes or embankment seawalls have been built along many Dutch, Danish or German coastlines protecting the land behind from flooding, and sometimes providing additional amenity value. Similar such structures in UK may alternatively be formed by clay materials or from a vegetated shingle ridge, in both instances allowing the side slopes to be steeper. All such embankments will need some degree of protection against direct wave erosion, generally using a revetment facing on the seaward side. Revetment facing may take many forms, but may commonly include closely-fitted concrete blockwork, cast in-situ concrete slabs, or asphaltic materials. Embankment or dike structures are generally most common along rural frontages.

A second type of coastal structure consists of a mound or layers of quarried rock fill, protected by rock or concrete armour units. The outer armour layer is designed to resist wave action without significant displacement of armour units. Under-layers of quarry or crushed rock support the armour and separate it from finer material in the embankment or mound. These porous and sloping layers dissipate a proportion of the incident wave energy in breaking and friction. Simplified forms of rubble mounds may be used for rubble seawalls or protection to vertical walls or revetments. Rubble mound revetments may also be used to protect embankments formed from relict sand dunes or shingle ridges. Rubble mound structures tend to be more common in areas where harder rock is available.

Along urban frontages, especially close to ports, erosion or flooding defence structures may include vertical (or battered/steep) walls. Such walls may be composed of stone or concrete blocks, mass concrete, or sheet steel piles. Typical vertical seawall structures may also act as retaining walls to material behind. Shaped and recurved wave return walls may be formed as walls in their own right, or smaller versions may be included in sloping structures. Some coastal structures are relatively impermeable to wave action. These include seawalls formed from blockwork or mass concrete, with vertical, near vertical, or steeply sloping faces. Such structures may be liable to intense local wave impact pressures, may overtop suddenly and severely, and will reflect much of the incident wave energy. Reflected waves cause additional wave disturbance and/or may initiate or accelerate local bed scour.

1.4 Definitions of key parameters and principal responses

Overtopping discharge occurs because of waves running up the face of a seawall. If wave run-up levels are high enough water will reach and pass over the crest of the wall. This defines the 'green water' overtopping case where a continuous sheet of water passes over the crest. In cases where the structure is vertical, the wave may impact against the wall and send a vertical plume of water over the crest.

A second form of overtopping occurs when waves break on the seaward face of the structure and produce significant volumes of splash. These droplets may then be carried over the wall either under their own momentum or as a consequence of an onshore wind.

Another less important method by which water may be carried over the crest is in the form of spray generated by the action of wind on the wave crests immediately offshore of the wall. Even with strong wind the volume is not large and this spray will not contribute to any significant overtopping volume.

Overtopping rates predicted by the various empirical formulae described within this report will include green water discharges and splash, since both these parameters were recorded during the model tests on which the prediction methods are based. The effect of wind on this type of discharge will not have been modelled. Model tests suggest that onshore winds have little effect on large green water events, however they may increase discharges under 1 l/s/m. Under these conditions, the water overtopping the structure is mainly spray and therefore the wind is strong enough to blow water droplets inshore.

In the list of symbols, short definitions of the parameters used have been included. Some definitions are so important that they are explained separately in this section as key parameters. The definitions and validity limits are specifically concerned with application of the given formulae. In this way, a structure section with a slope of 1:12 is not considered as a real slope (too gentle) and it is not a real berm too (too steep). In such a situation, wave run-up and overtopping can only be calculated by interpolation. For example, for a section with a slope of 1:12, interpolation can be made between a slope of 1:8 (mildest slope) and a 1:15 berm (steepest berm).

1.4.1 Wave height

The wave height used in the wave run-up and overtopping formulae is the incident significant wave height H_{m0} at the toe of the structure, called the spectral wave height, $H_{m0} = 4\,(m_0)^{½}$. Another definition of significant wave height is the average of the highest third of the waves, $H_{1/3}$. This wave height is, in principle, not used in this manual, unless formulae were derived on basis of it. In deep water, both definitions produce almost the same value, but situations in shallow water can lead to differences of 10–15 %.

In many cases, a foreshore is present on which waves can break and by which the significant wave height is reduced. There are models that in a relatively simple way can predict the reduction in energy from breaking of waves and thereby the accompanying wave height at the toe of the structure. The wave height must be calculated over the total spectrum including any long-wave energy present.

Based on the spectral significant wave height, it is reasonably simple to calculate a wave height distribution and accompanying significant wave height $H_{1/3}$ using the method of BATTJES and GROENENDIJK (2000).

1.4.2 Wave period

Various wave periods can be defined for a wave spectrum or wave record. Conventional wave periods are the peak period T_p (the period that gives the peak of the spectrum), the average period T_m (calculated from the spectrum or from the wave record) and the significant period $T_{1/3}$ (the average of the highest 1/3 of the waves). The relationship T_p/T_m usually lies between 1.1 and 1.25, and T_p and $T_{1/3}$ are almost identical.

The wave period used for some wave run-up and overtopping formulae is the spectral period $T_{m-1.0}$ $(= m_{-1}/m_0)$. This period gives more weight to the longer periods in the spectrum than an average period and, independent of the type of spectrum, gives similar wave run-up or overtopping for the same values of $T_{m-1.0}$ and the same wave heights. In this way, wave run-up and overtopping can be easily determined for double-peaked and ,flattened' spectra, without the need for other difficult procedures. Vertical and steep seawalls often use the $T_{m0,1}$ or T_m wave period.

In the case of a uniform (single peaked) spectrum there is a fixed relationship between the spectral period $T_{m-1.0}$ and the peak period. In this report a conversion factor ($T_p = 1.1\,T_{m-1.0}$) is given for the case where the peak period is known or has been determined, but not the spectral period.

1.4.3 Wave steepness and Breaker parameter

Wave steepness is defined as the ratio of wave height to wavelength (e.g. $s_0 = H_{m0}/L_0$). This will tell us something about the wave's history and characteristics. Generally a steepness of $s_0 = 0.01$ indicates a typical swell sea and a steepness of $s_0 = 0.04$ to 0.06 a typical wind sea. Swell seas will often be associated with long period waves, where it is the period that becomes the main parameter that affects overtopping.

But also wind seas may became seas with low wave steepness if the waves break on a gentle foreshore. By wave breaking the wave period does not change much, but the wave height decreases. This leads to a lower wave steepness. A low wave steepness on relatively deep water means swell waves, but for depth limited locations it often means broken waves on a (gentle) foreshore.

The breaker parameter, surf similarity or Iribarren number is defined as $\xi_{m-1,0} = \tan\alpha / (H_{m0}/L_{m-1,0})^{\frac{1}{2}}$, where α is the slope of the front face of the structure and $L_{m-1,0}$ being the deep water wave length $gT^2_{m-1,0}/2\pi$. The combination of structure slope and wave steepness gives a certain type of wave breaking, see Fig. 1.1. For $\xi_{m-1,0} > 2$–3 waves are considered not to be breaking (surging waves), although there may still be some breaking, and for $\xi_{m-1,0} < 2$–3 waves are breaking. Waves on a gentle foreshore break as spilling waves and more than one breaker line can be found on such a foreshore, see Fig. 1.2. Plunging waves break with steep and overhanging fronts and the wave tongue will hit the structure or back washing water; an example is shown in Fig. 1.3. The transition between plunging waves and surging waves is known as collapsing. The wave front becomes almost vertical and the water excursion on the slope (wave run-up + run down) is often largest for this kind of breaking. Values are given for the majority of the larger waves in a sea state. Individual waves may still surge for generally plunging conditions or plunge for generally surging conditions.

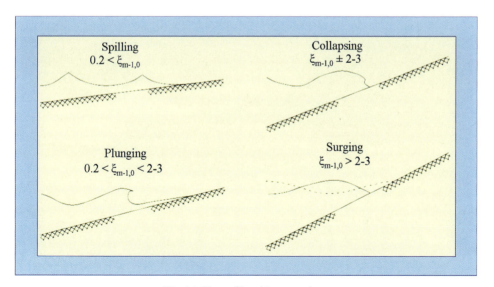

Fig. 1.1: Type of breaking on a slope

Fig. 1.2: Spilling waves on a beach; $\xi_{m-1,0} < 0.2$

Fig. 1.3: Plunging waves; $\xi_{m-1,0} < 2.0$

1.4.4 Parameter h*

In order to distinguish between non-impulsive (previously referred to as pulsating) waves on a vertical structure and impulsive (previously referred to as impacting) waves, the parameter h* has been defined.

$$h_* = \frac{h_s}{H_s} \frac{h_s}{L_o} \qquad \qquad 1.1$$

The parameter describes two ratios together, the wave height and wave length, both made relative to the local water depth h_s. Non-impulsive waves predominate when h* > 0.3; impulsive waves when h* ≤ 0.3. Formulae for impulsive overtopping on vertical structures, originally used this h* parameter to some power, both for the dimensionless wave overtopping and dimensionless crest freeboard.

1.4.5 Toe of structure

In most cases, it is clear where the toe of the structure lies, and that is where the foreshore meets the front slope of the structure or the toe structure in front of it. For vertical walls, it will be at the base of the principal wall, or if present, at the rubble mound toe in front of it. It is possible that a sandy foreshore varies with season and even under severe wave attack. Toe levels may therefore vary during a storm, with maximum levels of erosion occurring during the peak of the tidal/surge cycle. It may therefore be necessary to consider the effects of increased wave heights due to the increase in the toe depth. The wave height that is always used in wave overtopping calculations is the incident wave height at the toe.

1.4.6 Foreshore

The foreshore is the section in front of the dike and can be horizontal or up to a maximum slope of 1:10. The foreshore can be deep, shallow or very shallow. If the water is shallow or very shallow then shoaling and depth limiting effects will need to be considered so that the wave height at the toe; or end of the foreshore; can be considered. A foreshore is defined as having a minimum length of one wavelength L_o. In cases where a foreshore lies in very shallow depths and is relatively short, then the methods outlined in Section 5.3.4 should be used.

A precise transition from a shallow to a very shallow foreshore is hard to give. At a shallow foreshore waves break and the wave height decreases, but the wave spectrum will retain more or less the shape of the incident wave spectrum. At very shallow foreshores the spectral shape changes drastically and hardly any peak can be detected (flat spectrum). As the waves become very small due to breaking many different wave periods arise.

Generally speaking, the transition between shallow and very shallow foreshores can be indicated as the situation where the original incident wave height, due to breaking, has been decreased by 50 % or more. The wave height at a structure on a very shallow foreshore is much smaller than in deep water situations. This means that the wave steepness (Section 1.4.3) becomes much smaller, too. Consequently, the breaker parameter, which is used in the formulae for wave run-up and wave overtopping, becomes much larger. Values of ξ_o = 4 to 10 for the breaker parameter are then possible, where maximum values for a dike of 1:3 or 1:4 are normally smaller than say $\xi_0 = 2$ or 3.

Another possible way to look at the transition from shallow to very shallow foreshores, is to consider the breaker parameter. If the value of this parameter exceeds 5–7, or if they are swell waves, then a very shallow foreshore is present. In this way, no knowledge about wave heights at deeper water is required to distinguish between shallow and very shallow foreshores.

1.4.7 Slope

Part of a structure profile is defined as a slope if the slope of that part lies between 1:1 and 1:8. These limits are also valid for an average slope, which is the slope that occurs when a line is drawn between $-1.5\ H_{m0}$ and $+Ru_{2\%}$ in relation to the still water line and berms are not included. A continuous slope with a slope between 1:8 and 1:10 can be calculated in the first instance using the formulae for simple slopes, but the reliability is less than for steeper slopes. In this case interpolation between a slope 1:8 and a berm 1:15 is not possible.

A structure slope steeper than 1:1, but not vertical, can be considered as a battered wall. These are treated in Chapter 7 as a complete structure. If it is only a wave wall on top of gentle sloping dike, it is treated in Chapter 5.

1.4.8 Berm

A berm is part of a structure profile in which the slope varies between horizontal and 1:15. The position of the berm in relation to the still water line is determined by the depth, dh, the vertical distance between the middle of the berm and the still water line. The width of a berm, B, may not be greater than one-quarter of a wavelength, i.e., $B < 0.25\ L_o$. If the width is greater, then the structure part is considered between that of a berm and a foreshore, and wave run-up and overtopping can be calculated by interpolation. Section 5.3.4 gives a more detailed description.

1.4.9 Crest freeboard and armour freeboard and width

The crest height of a structure is defined as the crest freeboard, R_c, and has to be used for wave overtopping calculations. It is actually the point on the structure where overtopping water can no longer flow back to the seaside. The height (freeboard) is related to SWL. For rubble mound structures, it is often the top of a crest element and not the height of the rubble mound armour.

The armour freeboard, A_c, is the height of a horizontal part of the crest, measured relative to SWL. The horizontal part of the crest is called G_c. For rubble mound slopes the armour freeboard, A_c, may be higher, equal or sometimes lower than the crest freeboard, R_c, Fig. 1.4.

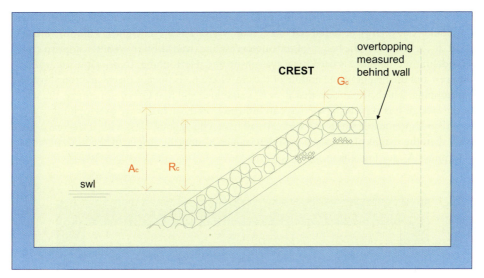

Fig. 1.4: Crest freeboard different from armour freeboard

The crest height that must be taken into account during calculations for wave overtopping for an upper slope with quarry stone, but without a wave wall, is not the upper side of this quarry stone, A_c. The quarry stone armour layer is itself completely water permeable, so that the under side must be used instead, see Fig. 1.5. In fact, the height of a non or only slightly water-permeable layer determines the crest freeboard, R_c, in this case for calculations of wave overtopping.

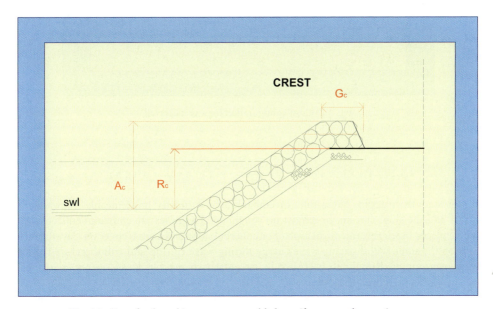

Fig. 1.5: Crest freeboard ignores a permeable layer if no crest element is present

The crest of a dike, especially if a road runs along it, is in many cases not completely horizontal, but slightly rounded and of a certain width. The crest height at a dike or embankment, R_c, is defined as the height of the outer crest line (transition from outer slope to crest). This definition therefore is used for wave run-up and overtopping. In principle the width of the crest and the height of the middle of the crest have no influence on calculations for wave overtopping, which also means that $R_c = A_c$ is assumed and that $G_c = 0$. Of course, the width of the crest, if it is very wide, can have an influence on the actual wave overtopping.

If an impermeable slope or a vertical wall have a horizontal crest with at the rear a wave wall, then the height of the wave wall determines R_c and the height of the horizontal part determines A_c, see Fig. 1.6.

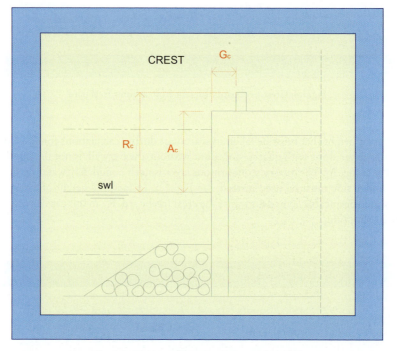

Fig. 1.6: Crest configuration for a vertical wall

1.4.10 Permeability, porosity and roughness

A smooth structure like a dike or embankment is mostly impermeable for water or waves and the slope has no, or almost no roughness. Examples are embankments covered with a placed block revetment, an asphalt or concrete slope and a grass cover on clay. Roughness on the slope will dissipate wave energy during wave run-up and will therefore reduce wave overtopping. Roughness is created by irregularly shaped block revetments or artificial ribs or blocks on a smooth slope.

A rubble mound slope with rock or concrete armour is also rough and in general more rough than roughness on impermeable dikes or embankments. But there is another difference, as the permeability and porosity is much larger for a rubble mound structure. Porosity is defined as the percentage of voids between the units or particles. Actually, loose materials always have some porosity. For rock and concrete armour the porosity may range roughly between 30 %–55 %. But also sand has a comparable porosity. Still the behaviour of waves on a sand beach or a rubble mound slope is different.

This difference is caused by the difference in permeability. The armour of rubble mound slopes is very permeable and waves will easily penetrate between the armour units and dissipate energy. But this becomes more difficult for the under layer and certainly for the core of the structure. Difference is made between "impermeable under layers or core" and a "permeable core". In both cases the same armour layer is present, but the structure and under layers differ.

A rubble mound breakwater often has an under layer of large rock (about one tenth of the weight of the armour), sometimes a second under layer of smaller rock and then the core of still smaller rock. Up-rushing waves can penetrate into the armour layer and will then sink into the under layers and core. This is called a structure with a "permeable core".

An embankment can also be covered by an armour layer of rock. The under layer is often small and thin and placed on a geotextile. Underneath the geotextile sand or clay may be present, which is impermeable for up-rushing waves. Such an embankment covered with rock has an "impermeable core". Run-up and wave overtopping are dependent on the permeability of the core.

In summary the following types of structures can be described:

Smooth dikes and embankments:	smooth and impermeable
Dikes and embankments with rough slopes:	some roughness and impermeable
Rock cover on an embankment:	rough with impermeable core
Rubble mound breakwater:	rough with permeable core

1.4.11 Wave run-up height

The wave run-up height is given by $R_{u2\%}$. This is the wave run-up level, measured vertically from the still water line, which is exceeded by 2 % of the number of incident waves. The number of waves exceeding this level is hereby related to the number of incoming waves and not to the number that run-up.

A very thin water layer in a run-up tongue cannot be measured accurately. In model studies on smooth slopes the limit is often reached at a water layer thickness of 2 mm. For prototype waves this means a layer depth of about 2 cm, depending on the scale in relation to the model study. Very thin layers on a smooth slope can be blown a long way up the slope by a strong wind, a condition that can also not be simulated in a small scale model. Running-up water tongues less than 2 cm thickness actually contain very little water. Therefore it is suggested that the wave run-up level on smooth slopes is determined by the level at which the water tongue becomes less than 2 cm thick. Thin layers blown onto the slope are not seen as wave run-up.

Run-up is relevant for smooth slopes and embankments and sometimes for rough slopes armoured with rock or concrete armour. Wave run-up is not an issue for vertical structures. The percentage or number of overtopping waves, however, is relevant for each type of structure.

1.4.12 Wave overtopping discharge

Wave overtopping is the mean discharge per linear meter of width, q, for example in $m^3/s/m$ or in $l/s/m$. The methods described in this manual calculate all overtopping discharges in $m^3/s/m$ unless otherwise stated; it is, however, often more convenient to multiply by 1000 and quote the discharge in $l/s/m$.

In reality, there is no constant discharge over the crest of a structure during overtopping. The process of wave overtopping is very random in time and volume. The highest waves will push a large amount of water over the crest in a short period of time, less than a wave period. Lower waves will not produce any overtopping. An example of wave overtopping measurements is shown in Fig. 1.7. The graphs shows 200 s of measurements. The lowest graph (flow depths) clearly shows the irregularity of wave overtopping. The upper graph gives the cumulative overtopping as it was measured in the overtopping tank. Individual overtopping volumes can be distinguished, unless a few overtopping waves come in one wave group.

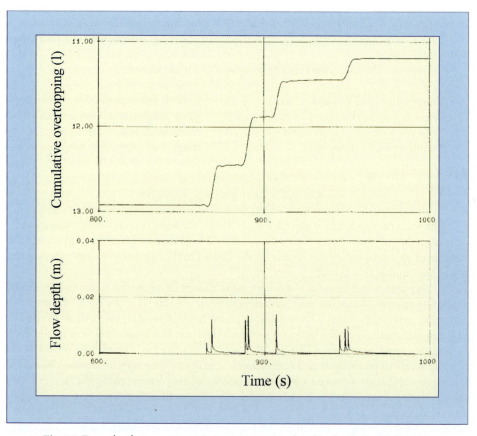

Fig. 1.7: Example of wave overtopping measurements, showing the random behaviour

Still a mean overtopping discharge is widely used as it can easily be measured and also classified:

q < 0.1 l/s per m: Insignificant with respect to strength of crest and rear of structure.
q = 1 l/s per m: On crest and inner slopes grass and/or clay may start to erode.
q = 10 l/s per m: Significant overtopping for dikes and embankments. Some overtopping for rubble mound breakwaters.
q = 100 l/s per m: Crest and inner slopes of dikes have to be protected by asphalt or concrete; for rubble mound breakwaters transmitted waves may be generated.

1.4.13 Wave overtopping volumes

A mean overtopping discharge does not yet describe how many waves will overtop and how much water will be overtopped in each wave. The volume of water, V, that comes over the crest of a structure is given in m³ per wave per m width. Generally, most of the overtopping waves are fairly small, but a small number gives significantly larger overtopping volumes.

The maximum volume overtopped in a sea state depends on the mean discharge q, on the storm duration and the percentage of overtopping waves. In this report, a method is given by which the distribution of overtopping volumes can be calculated for each wave. A longer storm duration gives more overtopping waves, but statistically, also a larger maximum volume. Many small overtopping waves (like for river dikes or embankments) may create the same mean overtopping discharge as a few large waves for rough sea conditions. The maximum volume will, however, be much larger for rough sea conditions with large waves.

1.5 Probability levels and uncertainties

This section will briefly introduce the concept of uncertainties and how it will be dealt with in this manual. It will start with a basic definition of uncertainty and return period. After that the various types of uncertainties are explained and more detailed descriptions of parameters and model uncertainties used in this manual will be described.

1.5.1 Definitions

Uncertainty may be defined as the relative variation in parameters or error in the model description so that there is no single value describing this parameter but a range of possible values. Due to the random nature of many of those variables used in coastal engineering most of the parameters should not be treated deterministically but stochastically. The latter assumes that a parameter x shows different realisations out of a range of possible values. Hence, uncertainty may be defined as a statistical distribution of the parameter. If a normal distribution is assumed here uncertainty may also be given as relative error, mathematically expressed as the coefficient of variation of a certain parameter x:

$$\sigma'_x = \frac{\sigma_x}{\mu_x} \qquad 1.2$$

where σ_x is the standard deviation of the parameter and μ_x is the mean value of that parameter. Although this definition may be regarded as imperfect it has some practical value and is easily applied.

The return period of a parameter is defined as the period of time in which the parameter occurs again on average. Therefore, it is the inverse of the probability of occurrence of this parameter. If the return period T_R of a certain wave height is given, it means that this specific wave height will only occur once in T_R years on average.

It should be remembered that there will not be exactly T_R years between events with a given return period of T_R years. If the events are statistically independent then the probability that a condition with a return period of T_R years will occur within a period of L years is given by $p = 1-(1-1/nT_R)^{nL}$, where n is the number of events per year, e.g., 2920 storms of three hours duration. Hence, for an event with a return period of 100 years there is a 1 % chance of recurrence in any one year. For a time interval equal to the return period, $p = 1-(1-1/nTr)^{nTr}$ or $p \approx 1 - 1/e = 0.63$. Therefore, there is a 63 % chance of occurrence within the return period. Further information on design events and return periods can be found in the British Standard Code of practice for Maritime Structures (BS6349 Part 1 1984 and Part 7 1991) or the PIANC working group 12 report (PIANC, 1992). Also refer to Section 2.6.

1.5.2 Background

Many parameters used in engineering models are uncertain, and so are the models themselves. The uncertainties of input parameters and models generally fall into certain categories; as summarised in Fig. 1.8.
- Fundamental or statistical uncertainties: elemental, inherent uncertainties, which are conditioned by random processes of nature and which can not be diminished (always comprised in measured data)
- Data uncertainty: measurement errors, inhomogeneity of data, errors during data handling, non-representative reproduction of measurement due to inadequate temporal and spatial resolution
- Model uncertainty: coverage of inadequate reproduction of physical processes in nature
- Human errors: all of the errors during production, abrasion, maintenance as well as other human mistakes which are not covered by the model. These errors are not considered in the following, due to the fact that in general they are specific to the problems and no universal approaches are available.

If Normal or Gaussian Distributions for x are used 68.3 % of all values of x are within the range of $\mu_x(1 \pm \sigma_x)$, 95.5 % of all values within the range of $\mu_x \pm 2\sigma_x$ and almost all values (97,7 %) within the range of $\mu_x \pm 3\sigma_x$, see Fig. 1.9. Considering uncertainties in a design, therefore, means that all input parameters are no longer regarded as fixed deterministic parameters but can be any realisation of the specific parameter. This has two consequences: Firstly, the parameters have to be checked whether all realisations of this parameter are really physically sound: E.g., a realisation of a normally distributed wave height can mathematically become negative which is physically impossible. Secondly, parameters have to be checked against realisations of other parameters: E.g., a wave of a certain height can only exist in certain water depths and not all combinations of wave heights and wave periods can physically exist.

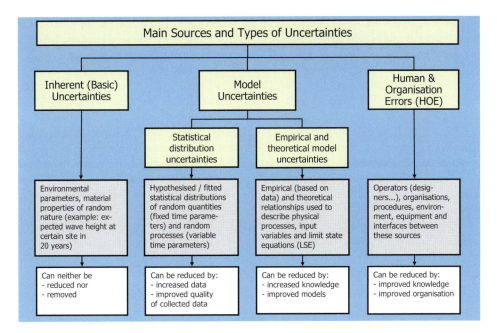

Fig. 1.8: Sources of uncertainties

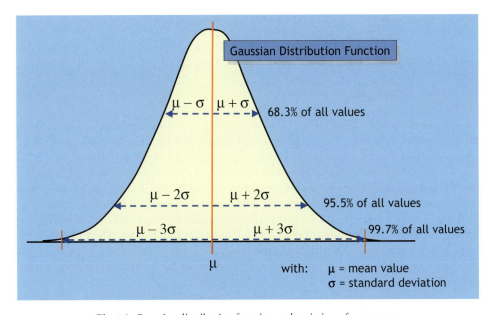

Fig. 1.9: Gaussian distribution function and variation of parameters

In designing with uncertainties this means that statistical distributions for most of the parameters have to be selected extremely carefully. Furthermore, physical relations between parameters have to be respected. This will be discussed in the subsequent sections as well.

1.5.3 Parameter uncertainty

The uncertainty of input parameters describes the inaccuracy of these parameters, either from measurements of those or from their inherent uncertainties. As previously discussed, this uncertainty will be described using statistical distributions or relative variation of these parameters. Relative variation for most of the parameters will be taken from various sources such as: measurement errors observed; expert opinions derived from questionnaires; errors reported in literature.

Uncertainties of parameters will be discussed in the subsections of each of the following chapters discussing various methods to predict wave overtopping of coastal structures. Any physical relations between parameters will be discussed and restrictions for assessing the uncertainties will be proposed.

1.5.4 Model uncertainty

The model uncertainty is considered as the accuracy, with which a model or method can describe a physical process or a limit state function. Therefore, the model uncertainty describes the deviation of the prediction from the measured data due to this method. Difficulties of this definition arise from the combination of parameter uncertainty and model uncertainty. Differences between predictions and data observations may result from either uncertainties of the input parameters or model uncertainty.

Model uncertainties may be described using the same approach than for parameter uncertainties using a multiplicative approach. This means that

$$q = m \cdot f(x_i) \qquad 1.3$$

where m is the model factor [–]; q is the overtopping ratio and f(x) is the model used for prediction of overtopping. The model factor m is assumed to be normally distributed with a mean value of 1.0 and a coefficient of variation specifically derived for the model.

These model factors may easily reach coefficients of variations up to 30 %. It should be noted that a mean value of m = 1.0 always means that there is no bias in the models used. Any systematic error needs to be adjusted by the model itself. For example, if there is an over-prediction of a specific model by 20 % the model has to be adjusted to predict 20 % lower results. This concept is followed in all further chapters of this manual so that from here onwards, the term 'model uncertainties' is used to describe the coefficient of variation σ', assuming that the mean value is always 1.0. The procedure to account for the model uncertainties is given in section 4.9.1.

Model uncertainties will be more widely discussed in the subsections of each chapter describing the models. The subsections will also give details on how the uncertain results of the specific model may be interpreted.

1.5.5 Methodology and output

All parameter and model uncertainties as discussed in the previous sections are used to run the models proposed in this manual. Results of all models will again follow statistical distributions rather than being single deterministic values. Hence, interpretation of these results is required and recommendations will be given on how to use outputs of the models.

Key models for overtopping will be calculated using all uncertainties and applying a Monte-Carlo-simulation (MCS). Statistical distributions of outputs will be classified with regard to exceedance probabilities such as: very safe, where output is only exceeded by 2 % of all results, corresponding to a return period of 50 years which means that the structure is expected to be overtopped only once during its lifetime of 50 years; safe, where output is exceeded by 10 % of all results, corresponding to a return period of 10 years; medium safe, where output corresponds to mean values plus one standard deviation; and probabilistic, where output is exceeded by 50 % of all results and may be used for probabilistic calculations.

2. Water levels and wave conditions

2.1 Introduction

This Overtopping Manual has a focus on the aspects of wave run-up and wave overtopping only. It is not a design manual, giving the whole design process of a structure. This chapter, therefore, will not provide a guide to the derivation of input conditions other than to identify the key activities in deriving water level and wave conditions, and particularly depth-limited wave conditions. It identifies the key parameters and provides a check-list of key processes and transformations. Comprehensive references are given to appropriate sources of information. Brief descriptions of methods are sometimes given, summary details of appropriate tools and models, and cross references to other manuals.

The main manuals and guidelines, which describe the whole design process of coastal and inland structures, including water levels and wave conditions are: The Rock Manual (1991), recently replaced by the updated Rock Manual (2007); The Coastal Engineering Manual; The British Standards; The German "Die Küste" (2002) and the DELOS Design Guidelines (2007).

2.2 Water levels, tides, surges and sea level changes

Prediction of water levels is extremely important for prediction of wave run-up levels or wave overtopping, which are often used to design the required crest level of a flood defence structure or breakwater. Moreover, in shallow areas the extreme water level often determines the water depth and thereby the upper limit for wave heights.

Extreme water levels in design or assessment of structures may have the following components: the mean sea level; the astronomical tide; surges related to (extreme) weather conditions; and high river discharges

2.2.1 Mean sea level

For coastal waters in open communication with the sea, the mean water level can often effectively be taken as a site-specific constant, being related to the mean sea level of the oceans. For safety assessments, not looking further ahead than about 5 years, the actual mean water level can be taken as a constant. Due to expected global warming, however, predictions in sea level rise for the next hundred years range roughly from 0.2 m to more than 1.0 m.

For design of structures, which last a long time after their design and construction phase, a certain sea level rise has to be included. Sometimes countries prescribe a certain sea level rise, which has to be taken into account when designing flood defence structures. Also the return period to include sea level rise may differ, due to the possibility of modification in future. An earthen dike is easy to increase in height and a predicted sea level rise for the next 50 years would be sufficient. A dedicated flood defence structure through a city is not easy to modify or replace. In such a situation a predicted sea level rise for the next 100 years or more could be considered.

2.2.2 Astronomical tide

The basic driving forces of tidal movements are astronomical and therefore entirely predictable, which enables accurate prediction of tidal levels (and currents). Around the UK and North Sea coast, and indeed around much of the world, the largest fluctuations in water level are caused by astronomical tides. These are caused by the relative rotation of both the sun and the moon around the earth each day. The differential gravitational effects over the surface of the oceans cause tides with well defined periods, principally semi-diurnal and diurnal. Around the British Isles and along coasts around the North Sea the semi-diurnal tides are much larger than the diurnal components.

In addition to the tides that result from the earth's rotation, other periodicities are apparent in the fluctuation of tidal levels. The most obvious is the fortnightly spring-neap cycle, corresponding to the half period of the lunar cycle.

Further details on the generation of astronomic tides, and their dynamics, can be found in the Admiralty Manual of Tides in most countries. These give daily predictions of times of high and low waters at selected locations, such as ports. Also details of calculating the differences in level between different locations are provided. Unfortunately, in practice, the prediction of an extreme water level is made much more complicated by the effects of weather, as discussed below.

2.2.3 Surges related to extreme weather conditions

Generally speaking the difference between the level of highest astronomical tide and, say, the largest predicted tide in any year is rather small (i.e. a few centimetres). In practice, this difference is often unimportant, when compared with the differences between predicted and observed tidal levels due to weather effects.

Extreme high water levels are caused by a combination of high tidal elevations plus a positive surge, which usually comprise three main components. A barometric effect caused by a variation in atmospheric pressure from its mean value. A wind set-up; in shallow seas, such as a the English Channel or the North Sea, a strong wind can cause a noticeable rise in sea level within a few hours. A dynamic effect due to the amplification of surge-induced motions caused by the shape of the land (e.g. seiching and funnelling).

A fourth component, wave set-up causes an increase in water levels within the surf zone at a particular site due to waves breaking as they travel shoreward. Unlike the other three positive surge components, wave set-up has only an extremely localised effect on water levels. Wave set-up is implicitly reproduced in the physical model tests on which the overtopping equations are based. There is, therefore, no requirement to add on an additional water level increase for wave set-up when calculating overtopping discharges using the methods reported in this document.

Negative surges are made up of two principal components: a barometric effect caused by high atmospheric pressures and wind set-down caused by winds blowing offshore. Large positive surges are more frequent than large negative ones. This is because a depression causing a positive surge will tend to be more intense and associated with a more severe wind condition than anticyclones.

Surges in relatively large and shallow areas, like the southern part of the North Sea, play an important role in estimating extreme water levels. The surges may become several meters for large return periods. The easiest means of predicting extreme water levels is to analyse long term water level data from the site in question. However, where no such data exists, it may be necessary to predict surge levels using theoretical equations and combine these levels with tidal elevations in order to obtain an estimation of extreme water levels.

More than 100 years' of high water level measurements in the Netherlands is shown in Fig. 2.1 along with the extrapolation of the measurements to extreme low exceedance probabilities, such as 10^{-4} or only once in 10,000 years.

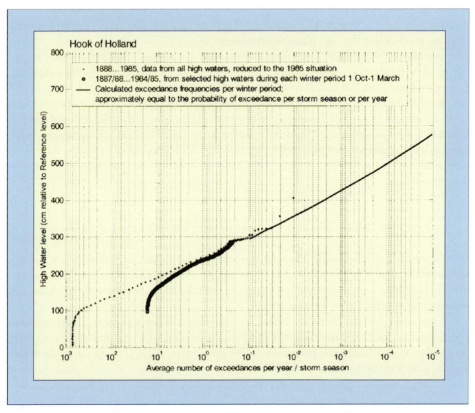

Fig. 2.1: Measurements of maximum water levels for more than 100 years and extrapolation to extreme return periods

2.2.4 High river discharges

Coastal flood defences face the sea or a (large) lake, but flood defences are also present along tidal rivers. Extreme river discharges determine the extreme water levels along river flood defences. During such an extreme water level, which may take a week or longer, a storm may generate waves on the river and cause overtopping of the flood defence. In many cases the required height of a river dike does not only depend on the extreme water level, but also on the possibility of wave overtopping. It should be noted that the occurrence of the extreme river discharge, and extreme water level, are independent of the occurrence of the storm. During high river discharges, only "normal" storms; occurring every decade; are considered, not the extreme storms.

Where rivers enter the sea both systems for extreme water levels may occur. Extreme storms may give extreme water levels, but also extreme river discharges. The effect of extreme storms and surges disappear farther upstream. Joint probabilistic calculations of both phenomena may give the right extreme water levels for design or safety assessment.

2.2.5 Effect on crest levels

During design or safety assessment of a dike, the crest height does not just depend on wave run-up or wave overtopping. Account must also be taken of a reference level, local sudden gusts and oscillations (leading to a corrected water level), settlement and an increase of the water level due to sea level rise.

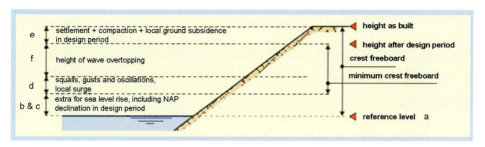

Fig. 2.2: Important aspects during calculation or assessment of dike height

The structure height of a dike in the Netherlands is composed of the following contributions; see also the Guidelines for Sea and Lake Dikes [TAW, 1999-2]:
a) the reference level with a probability of being exceeded corresponding to the legal standard (in the Netherlands this is a return period between 1,250 and 10,000 years;
b) the sea level rise or lake level increase during the design period;
c) the expected local ground subsidence during the design period;
d) an extra due to squalls, gusts, seiches and other local wind conditions;
e) the expected decrease in crest height due to settlement of the dike body and the foundation soils during the design period;
f) the wave run-up height and the wave overtopping height.

Contributions (a) to (d) cannot be influenced, whereas contribution (e) can be influenced. Contribution (f) also depends on the outer slope, which can consist of various materials, such as an asphalt layer, a cement-concrete dike covering (stone setting) or grass on a clay layer. A combination of these types is also possible. Slopes are not always straight, and the upper and lower sections may have different slopes and also a berm may be applied. The design of a covering layer is not dealt with in this report. However, the aspects related to berms, slopes and roughness elements are dealt with when they have an influence on wave run-up and wave overtopping.

2.3 Wave conditions

In defining the wave climate at the site, the ideal situation is to collect long term instrumentally measured data at the required location. There are very few instances in which this is even a remote possibility. The data of almost 30 years' of wave height measurements is shown in Fig. 2.3. These are the Dutch part of the North Sea with an extrapolation to very extreme events.

It is however more likely that data in deep water, offshore of a site will be available either through the use of a computational wave prediction model based on wind data, or on a wave model. In both of these cases the offshore data can be used in conjunction with a wave transformation model to provide information on wave climate at a coastal site. If instrumentally measured data is also available, covering a short period of time, this can be used for the calibration or verification of the wave transformation model, thus giving greater confidence in its use.

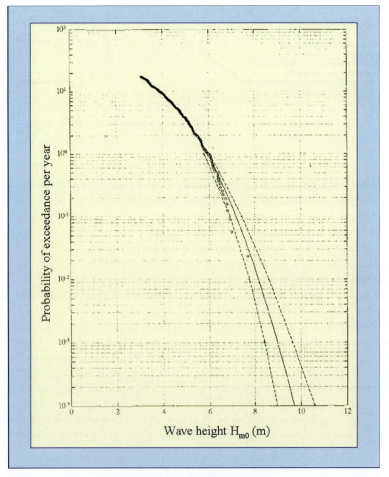

Fig. 2.3: Wave measurements and numerical simulations in the North Sea (1964–1993), leading to an extreme distribution

Wind generated waves offshore of most coasts have wave periods in the range 1s to 20s. The height, period and direction of the waves generated will depend on the wind speed, duration, direction and the 'fetch', i.e. the unobstructed distance of sea surface over which the wind has acted. In most situations, one of either the duration or fetch become relatively unimportant. For example, in an inland reservoir or lake, even a short storm will produce large wave heights. However, any increase in the duration of the wind will then cause no extra growth because of the small fetch lengths. Thus such waves are described as 'fetch limited'. In contrast, on an open coast where the fetch is very large but the wind blows for only a short period, the waves are limited by the duration of the storm. Beyond a certain limit, the exact fetch length becomes unimportant. These waves are described as 'duration limited'.

On oceanic shorelines the situation is usually more complicated. Both the fetch and duration may be extremely large, waves then become "fully developed" and their height depends solely on the wind speed. In such situations the wave period usually becomes quite large, and long period waves are able to travel great distances without suffering serious diminution. The arrival of 'swell', defined as waves not generated by local and/or recent wind conditions, presents a more challenging situation from the viewpoint of wave predictions.

2.4 Wave conditions at depth-limited situations

Wave breaking remains one phenomenon that is difficult to describe mathematically. One reason for this is that the physics of the process is not yet completely understood. However, as breaking has a significant effect on the behaviour of waves, the transport of sediments, the magnitude of forces on coastal structures and the overtopping response, it is represented in computational models. The most frequent method for doing this is to define an energy dissipation term which is used in the model when waves reach a limiting depth compared to their height.

There are also two relatively simple empirical methods for a first estimate of the incident wave conditions in the surf zone. The methods by GODA (1980) and OWEN (1980) are regularly used. GODA (1980) inshore wave conditions are influenced by shoaling and wave breaking. These processes are influenced by a number of parameters such as the sea steepness and the slope of the bathymetry. To take all the important parameters into account GODA (1980) provided a series of graphs to determine the largest and the significant wave heights (H_{max} and H_s) for 1:10, 1:20, 1:30 and 1:100 sloping bathymetries.

Results obtained from a simple 1D energy decay numerical model (VAN DER MEER, 1990) in which the influence of wave breaking is included, are presented in Fig. 2.4. This method has also been described in the Rock Manual (1991) and the updated version of this Rock Manual (2007). Tests have shown that wave height predictions using the design graphs from this model are accurate for slopes ranging from 1:10 to 1:100. For slopes flatter than 1:100, the predictions for the 1:100 slopes should be used.

The method for using these graphs is:
1. Determine the deep-water wave steepness, $s_{op} = H_{so}/L_{op}$ (where $L_{op} = gT_p^2/(2\pi)$). This value determines which graphs should be used. Suppose here for convenience that $s_{op} = 0.043$, then the graphs of Fig. 2.4 for $s_{op} = 0.04$ and 0.05 have to be used, interpolating between the results from each.
2. Determine the local relative water depth, h/L_{op}. The range of the curves in the graphs covers a decrease in wave height by 10 per cent to about 70 per cent. Limited breaking

occurs at the right hand side of the graphs and severe breaking on the left-hand side. If h/L_{op} is larger than the maximum value in the graph this means that there is no or only limited wave breaking and one can then assume no wave breaking (deep-water wave height = shallow-water wave height).
3. Determine the slope of the foreshore (m = tan α). Curves are given for range m = 0.075 to 0.01 (1:13 to 1:100). For gentler slopes the 1:100 slope should be used.
4. Enter the two selected graphs with calculated h/L_{op} and read the breaker index H_{m0}/h from the curve of the calculated foreshore slope.
5. Interpolate linearly between the two values of H_{m0}/h to find H_{m0}/h for the correct wave steepness.

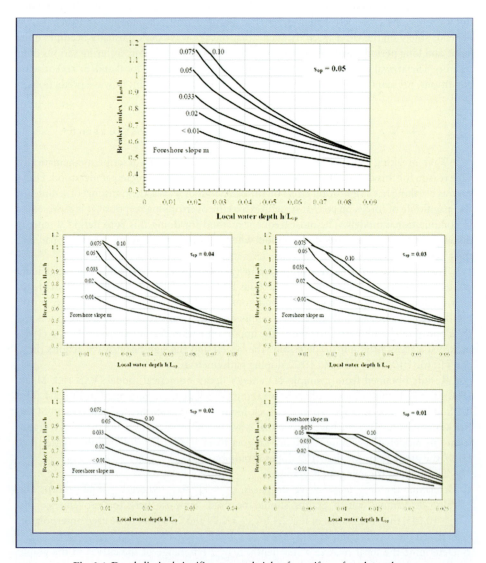

Fig. 2.4: Depth-limited significant wave heights for uniform foreshore slopes

Example. Suppose H_{so} = 6 m, T_p = 9.4 s, foreshore slope is 1:40 (m = 0.025). Calculate the maximum significant wave height H_{m0} at a water depth of h = 7 m.
1. The wave conditions on deep water give s_{op} = 0.043. Graphs with s_{op} = 0.04 and 0.05 have to be used.
2. The local relative water depth h/L_{op} = 0.051.
3. The slope of the foreshore (m = 0.025) is in between the curves for m = 0.02 and 0.033.
4. From the graphs, H_{m0}/h = 0.64 is found for s_{op} = 0.04 and 0.68 is found for s_{op} = 0.05.
5. Interpolation for s_{op} = 0.043 gives H_{m0}/h = 0.65 and finally a depth-limited spectral significant wave height of H_{m0} = 3.9 m.

Wave breaking in shallow water does not only affect the significant wave height H_{m0}. Also the distribution of wave heights will change. In deep water wave heights have a Rayleigh distribution and the spectral wave height H_{m0} will be close to the statistical wave height $H_{1/3}$. In shallow water these wave heights become different values due to the breaking process. Moreover, the highest waves break first when they feel the bottom, where the small waves stay unchanged. Actually, this gives a non-homogeneous set of wave heights: broken waves and non-broken waves. For this reason BATTJES and GROENENDIJK (2000) developed the composite Weibull distribution for wave heights in shallow water.

Although prediction methods in this manual are mainly based on the spectral significant wave height, it might be useful in some cases to consider also other definitions, like the 2%-wave height $H_{2\%}$ or $H_{1/10}$, the average of the highest 1/10-the of the waves. For this reason a summary of the method of BATTJES and GROENENDIJK (2000) is given here. The example given above with a calculated H_{m0} = 3.9 m at a depth of 7 m on a 1:40 slope foreshore has been explored further in Fig. 2.5.

$$H_{m0} = 4\sqrt{m_0} \quad H_{m0} = 4\sqrt{m_0}$$
$$H_{rms} = \left(2.69 + 3.24\sqrt{m_0}/h\right)\sqrt{m_0} \qquad 2.1$$

where H_{rms} = root mean square wave height. The transition wave height, H_{tr}, between the lower Rayleigh distribution and the higher Weibull distribution (see Fig. 2.5) is then given by:

$$H_r = (0.35 + 5.8 \tan \alpha)h \qquad 2.2$$

One has then to compute the non-dimensional wave height H_{tr}/H_{rms}, which is used as input to Table 2 of BATTJES and GROENENDIJK (2000) to find the (non-dimensional) characteristic heights: $H_{1/3}/H_{rms}$, $H_{1/10}/H_{rms}$, $H_{2\%}/H_{rms}$, $H_{1\%}/H_{rms}$ and $H_{0.1\%}/H_{rms}$. Some particular values have been extracted from this table and are included in Table 2.1, only for the ratios $H_{1/3}/H_{rms}$, $H_{1/10}/H_{rms}$, and $H_{2\%}/H_{rms}$.

Table 2.1: Values of dimensionless wave heights for some values of H_{tr}/H_{rms}

Characteristic height	Non-dimensional transitional wave H_{tr}/H_{rms}								
	0.05	0.50	1.00	1.20	1.35	1.50	1.75	2.00	2.50	3.00
$H_{1/3}/H_{rms}$	1.279	1.280	1.324	1.371	1.395	1.406	1.413	1.415	1.416	1.416
$H_{1/10}/H_{rms}$	1.466	1.467	1.518	1.573	1.626	1.683	1.759	1.786	1.799	1.800
$H_{2\%}/H_{rms}$	1.548	1.549	1.603	1.662	1.717	1.778	1.884	1.985	1.978	1.978

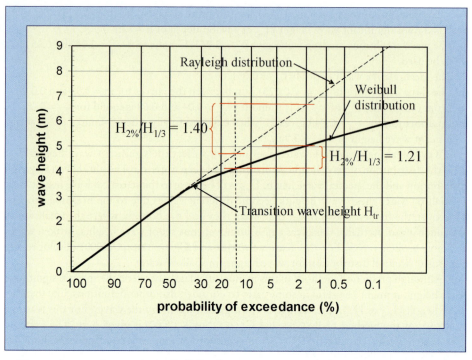

Fig. 2.5: Computed composite Weibull distribution. H_{m0} = 3.9 m; foreshore slope 1:40 and water depth h = 7 m

The final step is the computation of the dimensional wave heights from the ratios read in the table and the value of H_{rms}. For the given example one finds: $H_{1/3}$ = 4.16 m; $H_{1/10}$ = 4.77 m and $H_{2\%}$ = 5.4 m. Note that the value $H_{2\%}/H_{1/3}$ changed from 1.4 for a Rayleigh distribution (see Fig. 2.5) to a value of 1.21.

2.5 Currents

Where waves are propagating towards an oncoming current, for example at the mouth of a river, the current will tend to increase the steepness of the waves by increasing their height and decreasing their wave length. Refraction of the waves by the current will tend to focus the energy of the waves towards the river mouth. In reality both current and depth refraction are likely to take place producing a complex wave current field. It is clearly more complicated to include current and depth refraction effects, but at sites where currents are large they will have a significant influence on wave propagation. Computational models are available to allow both these effects to be represented.

2.6 Application of design conditions

The selection of a given return period for a particular site will depend on several factors. These will include the expected lifetime of the structure, expected maximum wave and water level conditions and the intended use of the structure. If for instance the public are to have access to the site then a higher standard of defence will be required than that to protect farm land. Further examples are given in Chapter 3.

A way of considering an event with a given return period, T_R, is to consider that (for $T_R \geq 5$ years) the probability of its occurrence in any one year is approximately equal to $1/T_R$. For example, a 10,000 year return period event is equivalent to one with a probability of occurrence of 10^{-4} in any one year.

Over an envisaged lifetime of N years for a structure (not necessarily the same as the design return period) the probability of encountering the wave condition with return period T_R, at least once, is given by:

$$P(\underline{T}_R \geq T_R) = 1 - (1 - 1/T_R)^N \qquad 2.3$$

Fig. 2.6 presents curves for this encounter probability with values between 1 per cent and 80 per cent shown as a function of T_R and N. It follows that there will not be exactly T_R years between events with a given return period of T_R years. It can be seen that for a time interval equal to the return period, there is a 63 % chance of occurrence within the return period. Further information on design events and return periods can be found in the British Standard Code of practice for Maritime Structures (BS6349 Part 1 1974 and Part 7 1991), the PIANC working group 12 report (PIANC 1992) and in the new Rock Manual (2007).

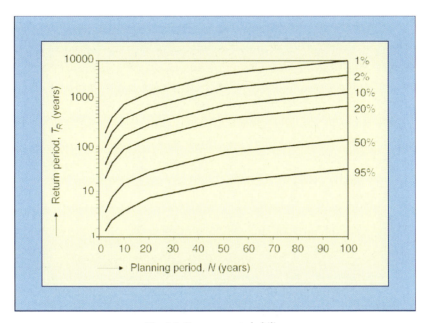

Fig. 2.6: Encounter probability

2.7 Uncertainties in inputs

Principal input parameters discussed in this section comprised water levels, including tides, surges, and sea level changes. Sea state parameters at the toe of the structure have been discussed and river discharges and currents have been considered.

It is assumed here that all input parameters are made available at the toe of the structure. Depending on different foreshore conditions and physical processes such as refraction, shoaling and wave breaking the statistical distributions of those parameters will have changed over the foreshore. Methods to account for this change are given in BATTJES & GROENENDIJK (2000) and elsewhere.

If no information on statistical distributions or error levels is available for water levels or sea state parameters the following assumptions should be taken: all parameters are normally distributed; significant wave height H_s or mean wave height H_{m0} have a coefficient of variation $\sigma_x' = 5.0\%$; peak wave period T_p or mean wave period $T_{m-1.0}$ have a coefficient of variation $\sigma_x' = 5.0\%$; and design water level at the toe $\sigma_x' = 3.0\%$, see SCHÜTTRUMPF et al. (2006).

The aforementioned values were derived from expert opinions on these uncertainties. About 100 international experts and professionals working in coastal engineering have been interviewed for this purpose. Although these parameters may be regarded rather small in relation to what GODA (1985) has suggested results have been tested against real cases and found to give a reasonable range of variations. It should be noted that these uncertainties are applied to significant values rather than mean sea state parameters. This will both change the type of the statistical distribution and the magnitude of the standard deviation or the coefficient of variation.

Guidance on hydraulic boundary conditions for the safety assessment of Dutch water defences can be found in Hydraulische Randvoorwaarden, RWS 2001 (Due to be updated in 2007).

3. Tolerable discharges

3.1 Introduction

Most sea defence structures are constructed primarily to limit overtopping volumes that might cause flooding. Over a storm or tide, the overtopping volumes that can be tolerated will be site specific as the volume of water that can be permitted will depend on the size and use of the receiving area, extent and magnitude of drainage ditches, damage versus inundation curves, and return period. Guidance on modelling inundation flows is being developed within Floodsite (FLOODSITE), but flooding volumes and flows, per se, are not distinguished further in this chapter.

For sea defences that protect people living, working or enjoying themselves, designers and owners of these defences must, however, also deal with potential direct hazards from overtopping. This requires that the level of hazard and its probability of occurrence be assessed, allowing appropriate action plans to be devised to ameliorate risks arising from overtopping.

The main hazards on or close to sea defence structures are of death, injury, property damage or disruption from direct wave impact or by drowning. On average, approximately 2–5 people are killed each year in each of UK and Italy through wave action, chiefly on seawalls and similar structures (although this rose to 11 in UK during 2005). It is often helpful to analyse direct wave and overtopping effects, and their consequences under four general categories:

a) Direct hazard of injury or death to people immediately behind the defence;
b) Damage to property, operation and/or infrastructure in the area defended, including loss of economic, environmental or other resource, or disruption to an economic activity or process;
c) Damage to defence structure(s), either short-term or longer-term, with the possibility of breaching and flooding;
d) Low depth flooding (inconvenient but not dangerous).

The character of overtopping flows or jets, and the hazards they cause, also depend upon the geometry of the structure and of the immediate hinterland behind the seawall crest, and the form of overtopping. For instance, rising ground behind the seawall may permit visibility of incoming waves, and will slow overtopping flows. Conversely, a defence that is elevated significantly above the land defended may obscure visibility of incoming waves, and post-overtopping flows may increase in speed rather than reduce. Hazards caused by overtopping therefore depend upon both the local topography and structures as well as on the direct overtopping characteristics.

It is not possible to give unambiguous or precise limits to tolerable overtopping for all conditions. Some guidance is, however, offered here on tolerable mean discharges and maximum overtopping volumes for a range of circumstances or uses, and on inundation flows and depths. These limits may be adopted or modified depending on the circumstances and uses of the site.

3.1.1 Wave overtopping processes and hazards

Hazards driven by overtopping can be linked to a number of simple direct flow parameters:

- mean overtopping discharge, q;
- individual and maximum overtopping volumes, V_i and V_{max};
- overtopping velocities over the crest or promenade, horizontally and vertically, v_{xc} and v_{zc} or v_{xp} and v_{zp};
- overtopping flow depth, again measured on crest or promenade, d_{xc} or d_{xp}.

Less direct responses (or similar responses, but farther back from the defence) may be used to assess the effects of overtopping, perhaps categorised by:

- overtopping falling distances, x_c;
- post-overtopping wave pressures (pulsating or impulsive), p_{qs} or p_{imp};
- post-overtopping flow depths, d_{xc} or d_{xp}; and horizontal velocities, v_{xc} or v_{xp}.

The main response to these hazards has most commonly been the construction of new defences, but responses should now always consider three options, in increasing order of intervention:

a) Move human activities away from the area subject to overtopping and/or flooding hazard, thus modifying the land use category and/or habitat status;
b) Accept hazard at a given probability (acceptable risk) by providing for temporary use and/or short-term evacuation with reliable forecast, warning and evacuation systems, and/or use of temporary/demountable defence systems;
c) Increase defence standard to reduce risk to (permanently) acceptable levels probably by enhancing the defence and/or reducing loadings.

For any structure expected to ameliorate wave overtopping, the crest level and/or the front face configuration will be dimensioned to give acceptable levels of wave overtopping under specified extreme conditions or combined conditions (e.g. water level and waves). Setting acceptable levels of overtopping depends on:

- the use of the defence structure itself;
- use of the land behind;
- national and/or local standards and administrative practice;
- economic and social basis for funding the defence.

Under most forms of wave attack, waves tend to break before or onto sloping embankments with the overtopping process being relatively gentle. Relatively few water levels and wave conditions may cause "impulsive" breaking where the overtopping flows are sudden and violent. Conversely, steeper, vertical or compound structures are more likely to experience intense local impulsive breaking, and may overtop violently and with greater velocities. The form of breaking will therefore influence the distribution of overtopping volumes and their velocities, both of which will impact on the hazards that they cause.

Additional hazards that are not dealt with here are those that arise from wave reflections, often associated with steep faced defences. Reflected waves increase wave disturbance, which may cause hazards to navigating or moored vessels; may increase waves along neighbouring frontages, and/or may initiate or accelerate local bed erosion thus increasing depth-limited wave heights (see section 2.4).

3.1.2 Types of overtopping

Wave overtopping which runs up the face of the seawall and over the crest in (relatively) complete sheets of water is often termed 'green water'. In contrast, 'white water' or spray overtopping tends to occur when waves break seaward of the defence structure or break onto its seaward face, producing non-continuous overtopping, and/or significant volumes of spray. Overtopping spray may be carried over the wall either under its own momentum, or assisted and/or driven by an onshore wind. Additional spray may also be generated by wind acting directly on wave crests, particularly when reflected waves interact with incoming waves to give severe local 'clapotii'. This type of spray is not classed as overtopping nor is it predicted by the methods described in this manual.

Without a strong onshore wind, spray will seldom contribute significantly to overtopping volumes, but may cause local hazards. Light spray may reduce visibility for driving, important on coastal highways, and will extend the spatial extent of salt spray effects such as damage to crops/vegetation, or deterioration of buildings. The effect of spray in reducing visibility on coastal highways (particularly when intermittent) can cause sudden loss of visibility in turn leading drivers to veer suddenly.

Effects of wind and generation of spray have not often been modelled. Some research studies have suggested that effects of onshore winds on large green water overtopping are small, but that overtopping under $q = 1$ l/s/m might increase by up to 4 times under strong winds, especially where much of the overtopping is as spray. Discharges between $q = 1$ to 0.1 l/s/m are however already greater than some discharge limits suggested for pedestrians or vehicles, suggesting that wind effects may influence overtopping at and near acceptable limits for these hazards.

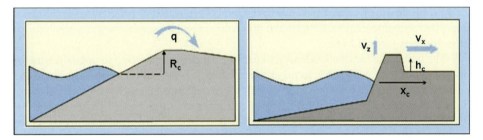

Fig. 3.1: Overtopping on embankment and promenade seawalls

3.1.3 Return periods

Return periods at which overtopping hazards are analysed, and against which a defence might be designed, may be set by national regulation or guidelines. As with any area of risk management, different levels of hazard are likely to be tolerated at inverse levels of probability or return period. The risk levels (probability x consequence) that can be tolerated will depend on local circumstances, local and national guidelines, the balance between risk and benefits, and the level of overall exposure. Heavily trafficked areas might therefore be designed to experience lower levels of hazard applied to more people than lightly used areas, or perhaps the same hazard level at longer return periods. Guidance on example return periods used in evaluating levels of protection suggest example protection levels versus return periods as shown in Table 3.1.

In practice, some of these return periods may be regarded as too short. National guidelines have recommended lower risk, e.g. a low probability of flooding in UK is now taken as <0.1 % probability (1:1000 year return) and medium probability of sea flooding as between

0.5 % and 0.1 % (1:200 to 1:1000 year return). Many existing sea defences in the UK however offer levels of protection far lower than these.

Table 3.1: Hazard Type

Hazard type and reason	Design life (years)	Level of Protection[1] (years)
Temporary or short term measures	1–20	5–50
Majority of coast protection or sea defence walls	30–70	50–100
Flood defences protecting large areas at risk	50–100	100–10,000
Special structure, high capital cost	200	Up to 10,000
Nuclear power stations etc.	–	10,000

[1] Note: Total probability return period

It is well known that the Netherlands is low-lying with two-thirds of the country below storm surge level. Levels of protection were increased after the flood in 1953 where almost 2000 people drowned. Large rural areas have a level of protection of 10,000 years, less densely populated areas a level of 4,000 years and protection for high river discharge (without threat of storm surge) of 1,250 years.

The design life for flood defences, like dikes, which are fairly easy to upgrade, is taken in the Netherlands as 50 years. In urban areas, where it is more difficult to upgrade a flood defence, the design life is taken as 100 years. This design life increases for very special structures with high capital costs, like the Eastern Scheldt storm surge barrier, Thames barrier, or the Maeslandtkering in the entrance to Rotterdam. A design life of around 200 years is then usual.

Variations from simple "acceptable risk" approach may be required for publicly funded defences based on benefit – cost assessments, or where public aversion to hazards causing death require greater efforts to ameliorate the risk, either by reducing the probability of the hazard or by reducing its consequence.

3.2 Tolerable mean discharges

Guidance on overtopping discharges that can cause damage to seawalls, buildings or infrastructure, or danger to pedestrians and vehicles have been related to mean overtopping discharges or (less often) to peak volumes. Guidance quoted previously were derived initially from analysis in Japan of overtopping perceived by port engineers to be safe (GODA et al. [1975], FUKUDA et al. [1974]). Further guidance from Iceland suggests that equipment or cargo might be damaged for $q \geq 0.4$ l/s/m. Significantly different limits are discussed for embankment seawalls with back slopes; or for promenade seawalls without back slopes. Some guidance distinguishes between pedestrians or vehicles, and between slow and faster speeds for vehicles.

Tests on the effects of overtopping on people suggest that information on mean discharges alone may not give reliable indicators of safety for some circumstances, and that maximum individual volumes may be better indicators of hazard than average discharges. The volume (and velocity) of the largest overtopping event can vary significantly with wave condition and structure type, even for a given mean discharge. There remain however two difficulties in specifying safety levels with reference to maximum volumes rather than to mean discharges. Methods to predict maximum volumes are available for fewer structure types, and are less well-validated. Secondly, data relating individual maximum overtopping volumes to hazard levels are still very rare.

In most instances the discharge (or volumes) discussed here are those at the point of interest, e.g. at the roadway or footpath or building. It is noted that the hazardous effect of overtopping waters reduces with the distance away from the defence line. As a rule of thumb, the hazard effect of an overtopping discharge at a point x metres back from the seawall crest will be to reduce the overtopping discharge by a factor of x, so the effective overtopping discharge at x (over a range of 5–25 m), $q_{effective}$ is given by:

$$q_{effective} = q_{seawall}/x. \qquad 3.1$$

The overtopping limits suggested in Table 3.2 to Table 3.5 therefore derive from a generally precautionary principle informed by previous guidance and by observations and measurements made by the CLASH partners and other researchers. Limits for pedestrians in Table 3.2 show a logical sequence, with allowable discharges reducing steadily as the recipient's ability or willingness to anticipate or receive the hazard reduces.

Table 3.2: Limits for overtopping for pedestrians

Hazard type and reason	Mean discharge q (l/s/m)	Max volume[1] V_{max} (l/m)
Trained staff, well shod and protected, expecting to get wet, overtopping flows at lower levels only, no falling jet, low danger of fall from walkway	1–10	500 at low level
Aware pedestrian, clear view of the sea, not easily upset or frightened, able to tolerate getting wet, wider walkway[2].	0.1	20–50 at high level or velocity

[1] Note: These limits relate to overtopping velocities well below $v_c \approx 10$ m/s. Lower volumes may be required if the overtopping process is violent and/or overtopping velocities are higher.
[2] Note: Not all of these conditions are required, nor should failure of one condition on its own require the use of a more severe limit.

A further precautionary limit of $q = 0.03$ l/s/m might apply for unusual conditions where pedestrians have no clear view of incoming waves; may be easily upset or frightened or are not dressed to get wet; may be on a narrow walkway or in close proximity to a trip or fall hazard. Research studies have however shown that this limit is only applicable for the conditions identified, and should NOT be used as the general limit for which $q = 0.1$ l/s/m in Table 3.2 is appropriate.

For vehicles, the suggested limits are rather more widely spaced as two very different situations are considered. The higher overtopping limit in Table 3.3 applies where wave overtopping generates pulsating flows at roadway level, akin to driving through slowly-varying fluvial flow across the road. The lower overtopping limit in Table 3.3 is however derived from considering more impulsive flows, overtopping at some height above the roadway, with overtopping volumes being projected at speed and with some suddenness. These lower limits are however based on few site data or tests, and may therefore be relatively pessimistic.

Table 3.3: Limits for overtopping for vehicles

Hazard type and reason	Mean discharge q (l/s/m)	Max volume V_{max} (l/m)
Driving at low speed, overtopping by pulsating flows at low flow depths, no falling jets, vehicle not immersed	10–50[1]	100–1,000
Driving at moderate or high speed, impulsive overtopping giving falling or high velocity jets	0.01–0.05[2]	5–50[2] at high level or velocity

[1] Note: These limits probably relate to overtopping defined at highway.
[2] Note: These limits relate to overtopping defined at the defence, but assumes the highway to be immediately behind the defence.

Rather fewer data are available on the effects of overtopping on structures, buildings and property. Site-specific studies suggest that pressures on buildings by overtopping flows will vary significantly with the form of wave overtopping, and with the use of sea defence elements intended to disrupt overtopping momentum (not necessarily reducing discharges). Guidance derived from the CLASH research project and previous work suggests limits in Table 3.4 for damage to buildings, equipment or vessels behind defences.

Table 3.4: Limits for overtopping for property behind the defence

Hazard type and reason	Mean discharge q (l/s/m)	Max volume V_{max} (l/m)
Significant damage or sinking of larger yachts	50	5,000–50,000
Sinking small boats set 5–10 m from wall. Damage to larger yachts	10[1]	1,000–10,000
Building structure elements	1[2]	~
Damage to equipment set back 5–10 m	0.4[1]	~

[1] Note: These limits relate to overtopping defined at the defence.
[2] Note: This limit relates to the effective overtopping defined at the building.

A set of limits for defence structures in Table 3.5 have been derived from early work by Goda and others in Japan. These give a first indication of the need for specific protection to resist heavy overtopping flows. It is assumed that any structure close to the sea will already be detailed to resist the erosive power of heavy rainfall and/or spray. Two situations are considered, see Fig. 3.1: Embankment seawalls or sea dikes with the defence structure elevated above the defended area, so overtopping flows can pass over the crest and down the rear face; or promenade defences in which overtopping flows remain on or behind the seawall crest before returning seaward. The limits for the latter category cannot be applied where the overtopping flows can fall from the defence crest where the nature of the flow may be more impulsive.

Table 3.5: Limits for overtopping for damage to the defence crest or rear slope

Hazard type and reason	Mean discharge q (l/s/m)
Embankment seawalls/sea dikes	
No damage if crest and rear slope are well protected	50–200
No damage to crest and rear face of grass covered embankment of clay	1–10
No damage to crest and rear face of embankment if not protected	0.1
Promenade or revetment seawalls	
Damage to paved or armoured promenade behind seawall	200
Damage to grassed or lightly protected promenade or reclamation cover	50

Wave overtopping tests were performed in early 2007 on a real dike in the Netherlands. The dike had a 1:3 inner slope of fairly good clay (sand content smaller than 30 %) with a grass cover. The wave overtopping simulator (see Section 3.3.3) was used to test the erosion resistance of this inner slope. Tests were performed simulating a 6 hour storm for every overtopping condition at a constant mean overtopping discharge. These conditions started with a mean discharge of 0.1 l/s/m and increased to 1; 10; 20; 30 and finally even 50 l/s/m. After all these simulated storms the slope was still in good condition and showed little erosion. The erosion resistance of this dike was very high.

Another test was performed on bare clay by removing the grass sod over the full inner slope to a depth of 0.2 m. Overtopping conditions of 0.1 l/s/m; 1; 5 and finally 10 l/s/m were performed, again for 6 hours each. Erosion damage started for the first condition (two erosion holes) and increased during the other overtopping conditions. After 6 hours at a mean discharge of 10 l/s/m (see Fig. 3.2) there were two large erosion holes, about 1 m deep, 1 m wide and 4 m long. This situation was considered as "not too far from initial breaching".

The overall conclusion of this first overtopping test on a real dike is that clay with grass can be highly erosion resistant. Even without grass the good quality clay also survived extensive overtopping. The conclusions may not yet be generalized to all dikes as clay quality and type of grass cover still may play a role and, therefore, more testing is required to come to general conclusions.

Fig. 3.2: Wave overtopping test on bare clay; result after 6 hours with 10 l/s per m width

One remark, however, should be made on the strength of the inner slopes of dikes by wave overtopping. Erosion of the slope is one of the possible failure mechanisms. The other one, which happened often in the past, is a slip failure of the slope. Slip failures may directly lead to a breach, but such slip failures occur mainly for steep inner slopes like 1:1.5 or 1:2. For this reason most dike designs in the Netherlands in the past fifty years have been based on a 1:3 inner slope, where it is unlikely that slip failures will occur due to overtopping. This mechanism might however occur for steep inner slopes, so should be taken into account in safety analysis.

3.3 Tolerable maximum volumes and velocities

3.3.1 Overtopping volumes

Guidance on suggested limits for maximum individual overtopping volumes have been given in Table 3.2 to Table 3.5 where data are available. Research studies with volunteers at full scale or field observations suggest that danger to people or vehicles might be related to peak overtopping volumes, with "safe" limits for people covering:

V_{max} = 1000 to 2000 l/m for trained and safety-equipped staff in pulsating flows on a wide-crested dike;
V_{max} = 750 l/m for untrained people in pulsating flows along a promenade;
V_{max} = 100 l/m for overtopping at a vertical wall
V_{max} = 50 l/m where overtopping could unbalance an individual by striking their upper body without warning.

3.3.2 Overtopping velocities

Few data are available on overtopping velocities and their contribution to hazards. For simply sloping embankments Chapter 5 gives guidance on overtopping flow velocities at crest and inner slope as well as on flow depths. Velocities of 5–8 m/s are possible for the maximum overtopping waves during overtopping discharges of about 10–30 l/s per m width. Studies of hazards under steady flows suggest that limits on horizontal velocities for people and vehicles will probably need to be set below v_x < 2.5 to 5 m/s. Also refer to Section 5.5.

Upward velocities (v_z) for vertical and battered walls under impulsive and pulsating conditions have been related to the inshore wave celerity, see Chapter 7. Relative velocities, v_z/c_i, have been found to be roughly constant at $v_z/c_i \approx 2.5$ for pulsating and slightly impulsive conditions, but increase significantly for impulsive conditions, reaching $v_z/c_i \approx 3 - 7$.

3.3.3 Overtopping loads and overtopping simulator

Post-overtopping wave loads have seldom been measured on defence structures, buildings behind sea defences, or on people, so little generic guidance is available. If loadings from overtopping flows could be important, they should be quantified by interpretation of appropriate field data or by site-specific model studies.

An example (site specific) model study indicates how important these effects might be. A simple 1 m high vertical secondary wall was set in a horizontal promenade about 7 m back from the primary seawall, itself a concrete recurve fronted by a rock armoured slope. Pulsating wave pressures were measured on the secondary wall against the effective overtopping discharge arriving at the secondary wall, plotted here in Fig. 3.3. This was deduced by applying Equation 3.1 to overtopping measured at the primary wall, 7 m in front. Whilst strongly site specific, these results suggest that quite low discharges (0.1–1.0 l/s/m) may lead to loadings up to 5kPa.

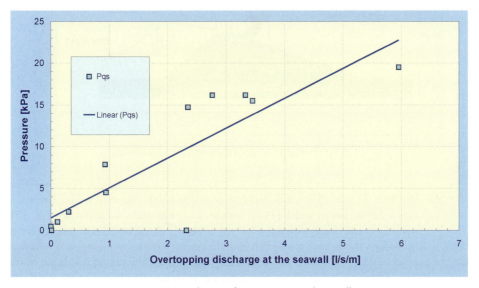

Fig. 3.3: Example wave forces on a secondary wall

During 2007, a new wave overtopping simulator was developed to test the erosion resistance of crest and inner slope of a dike, starting from the idea that:
- knowledge on wave breaking on slopes and overtopping discharges is sufficient (Chapter 5);
- knowledge on the pattern of overtopping volumes, distributions, velocities and flow depth of overtopping water on the crest, is sufficient as well (Chapter 5);
- only the overtopping part of the waves need to be simulated;
- tests can be performed in-situ on each specific dike, which is much cheaper than testing in a large wave flume.

The simulator was developed and designed within the ComCoast, see Fig. 3.4. Results of the calibration phase with a 1 m wide prototype were described by VAN DER MEER (2006).

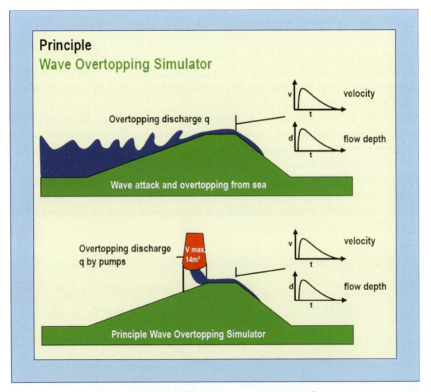

Fig. 3.4: Principle of the wave overtopping simulator

The simulator consists of a mobile box (adjustable in height) to store water. The maximum capacity is 3.5 m³ per m width (14 m³ for the final, 4 m wide, simulator see Fig. 3.5). This box is filled continuously with a predefined discharge q and emptied at specific times through a butterfly valve in such a way that it simulates the overtopping tongue of a wave at the crest and inner slope of a dike. As soon as the box contains the required volume, V, the valve is opened and the water is released on a transition section that leads to the crest of the dike. The discharge is released such that flow velocity, turbulence and thickness of the water

Fig. 3.5: The wave overtopping simulator discharging a large overtopping volume on the inner slope of a dike

tongue at the crest corresponds with the characteristics that can be expected (see Chapter 5). The calibration (VAN DER MEER, 2006) showed that it is possible to simulate the required velocities and flow depths for a wide range of overtopping rates, significantly exceeding present standards.

3.4 Effects of debris and sediment in overtopping flows

There are virtually no data on the effect of debris on hazards caused by overtopping, although anecdotal comments suggest that damage can be substantially increased for a given overtopping discharge or volume if "hard" objects such as rocks, shingle or timber are included in overtopping. It is known that impact damage can be particularly noticeable for seawalls and promenades where shingle may form the "debris" in heavy or frequent overtopping flows.

4. Prediction of overtopping

4.1 Introduction

A number of different methods may be available to predict overtopping of particular structures (usually simplified sections) under given wave conditions and water levels. Each method will have strengths or weaknesses in different circumstances. In theory, an analytical method can be used to relate the driving process (waves) and the structure to the response through equations based directly on a knowledge of the physics of the process. It is however extremely rare for the structure, the waves and the overtopping process to all be so simple and well-controlled that an analytical method on its own can give reliable predictions. Analytical methods are not therefore discussed further in this manual.

The primary prediction methods are therefore based on empirical methods (Section 4.2) that relate the overtopping response (usually mean overtopping discharge) to the main wave and structure parameters. Two other methods have been derived during the CLASH European project based on the use of measured overtopping from model tests and field measurements. The first of these techniques uses the CLASH database of structures, waves and overtopping discharges, with each test described by 13 parameters. Using the database (Section 4.5) is however potentially complicated, requiring some familiarity with these type of data. A simpler approach, and much more rapid, is to use the Neural Network tool (Section 4.3) that has been trained using the test results in the database. The Neural Network tool can be run automatically on a computer as a stand-alone device, or embedded within other simulation methods.

For situations for which empirical test data do not already exist, or where the methods above do not give reliable enough results, then two alternative methods may be used, but both are more complicated than the three methods described in Sections 4.2 to 4.5. A range of numerical models can be used to simulate the process of overtopping (Section 4.6). All such models involve some simplification of the overtopping process and are therefore limited to particular types of structure or types of wave exposure. They may however run sequences of waves giving overtopping (or not) on a wave-by-wave basis. Generally, numerical models require more skill and familiarity to run successfully.

The final method discussed here is physical modelling in which a scale model is tested with correctly scaled wave conditions. Typically such models may be built to a geometric scale typically in the range 1:20 to 1:60, see discussion on model and scale effects in Section 4.8. Waves will be generated as random wave trains each conforming to a particular energy spectrum. The model may represent a structure cross-section in a 2-dimensional model tested in a wave flume. Structures with more complex plan shapes, junctions, transitions etc., may be tested in a 3-dimensional model in a wave basin. Physical models can be used to measure many different aspects of overtopping such as wave-by-wave volumes, overtopping velocities and depths, as well as other responses.

4.2 Empirical models, including comparison of structures

4.2.1 Mean overtopping discharge

Empirical methods use a simplified representation of the physics of the process presented in (usually dimensionless) equations to relate the main response parameters (overtopping discharge etc) to key wave and structure parameters. The form and coefficients of the equations are adjusted to reproduce results from physical model (or field) measurements of waves and overtopping.

Empirical equations may be solved explicitly, or may occasionally require iterative methods to solve. Historically some empirical methods have been presented graphically, although this is now very rare.

The mean overtopping discharge, q, is the main parameter in the overtopping process. It is of course not the only parameter, but it is easy to measure in a laboratory wave flume or basin, and most other parameters are related in some way to this overtopping discharge. The overtopping discharge is given in m³/s per m width and in practical applications often in litres/s per m width. Although it is given as a discharge, the actual process of wave overtopping is much more dynamic. Only large waves will reach the crest of the structure and will overtop with a lot of water in a few seconds. This wave by wave overtopping is more difficult to measure in a laboratory than the mean overtopping discharge.

As the mean overtopping discharge is quite easy to measure many physical model tests have been performed all over the world, both for scientific (idealised) structures and real applications or designs. The European CLASH project resulted in a large database of more than 10,000 wave overtopping tests on all kind of structures (see Section 4.5). Some series of tests have been used to develop empirical methods for prediction of overtopping. Very often the empirical methods or formulae are applicable for typical structures only, like smooth slopes (dikes, sloping seawalls), rubble mound structures or vertical structures (caissons) or walls.

Chapters 5, 6 and 7 will describe in detail formulae for the different kinds of structure. In this section an overall view will be given in order to compare different structures and to give more insight into how wave overtopping behaves for different kind of structures. The structures considered here with governing overtopping equations (more details in Chapters 5, 6 and 7) are: smooth sloping structures (dikes, seawalls); rubble mound structures (breakwaters, rock slopes); and vertical structures (caissons, sheet pile walls).

The principal formula used for wave overtopping is:

$$\frac{q}{\sqrt{gH_{m0}^3}} = a\exp(-bR_c/H_{m0}) \qquad 4.1$$

It is an exponential function with the dimensionless overtopping discharge $q/(gH_{m0}^3)^{½}$ and the relative crest freeboard R_c/H_{m0}. This type of equation shown in a log-linear graph gives a straight line, which makes it easy to compare the formulae for various structures. Specific equations are given in Chapters 5 and 6 for smooth and rubble mound structures and sometimes include a berm, oblique wave attack, wave walls and the slope angle and wave period or wave steepness.

Two equations are considered for pulsating waves on a vertical structure. ALLSOP et al. (1995) consider relatively shallow water and FRANCO et al. (1994) more deep water (caissons). Vertical structures in shallow water, and often with a sloping foreshore in front, may become subject to impulsive forces, i.e. high impacts and water splashing high up into the air. Specific formulae have been developed for these kinds of situation.

For easy comparison of different structures, like smooth and rubble mound sloping structures and vertical structures for pulsating and impulsive waves, some simplifications will be assumed.

In order to simplify the smooth structure no berm is considered ($\gamma_b = 1$), only perpendicular wave attack is present ($\gamma_\beta = 1$), and no vertical wall on top of the structure is present

($\gamma_v = 1$). As a smooth structure is considered also, $\gamma_f = 1$. This limits the structure to a smooth and straight slope with perpendicular wave attack. The slope angles considered for smooth slopes are $\cot\alpha$ = 1 to 8, which means from very steep to very gentle. If relevant a wave steepness of $s_o = 0.04$ (steep storm waves) and 0.01 (long waves due to swell or wave breaking) will be considered.

The same equation as for smooth sloping structures is applicable for rubble mound slopes, but now with a roughness factor of $\gamma_f = 0.5$, simulating a rock structure. Rubble mound structures are often steep, but rock slopes may also be gentle. Therefore slope angles with $\cot\alpha = 1.5$ and 4.0 are considered.

For vertical structures under pulsating waves both formulae of ALLSOP et al. (1995) and FRANCO et al. (1994) will be compared, together with the formula for impulsive waves. Impulsive waves can only be reached with a relatively steep foreshore in front of the vertical wall. For comparison values of the ratio wave height/water depth of $H_{m0}/h_s = 0.5, 0.7$ & 0.9 will be used.

Smooth slopes can be compared with rubble mound slopes and with vertical structures under pulsating or impulsive conditions. First the traditional graph is given in Fig. 4.1 with the relative freeboard R_c/H_{m0} versus the logarithmic dimensionless overtopping $q/(gH_{m0}^3)^{1/2}$.

In most cases the steep smooth slope gives the largest overtopping. Steep means $\cot\alpha < 2$, but also a little gentler if long waves (small steepness) are considered. Under these conditions waves surge up the steep slope. For gentler slopes waves break as plunging waves and this reduces wave overtopping. The gentle slope with $\cot\alpha = 4$ gives much lower overtopping than the steep smooth slopes. Both slope angle and wave period have influence on overtopping for gentle slopes.

The large roughness and high permeability of a rubble mound structures reduces wave overtopping to a greater extent; see Fig. 4.1. A roughness factor of $\gamma_f = 0.5$ was used and a value of 0.4 (two layers of rock on a permeable under layer) would even reduce the overtopping further. The gentle rubble mound slope with $\cot\alpha = 4$ gives very low overtopping.

Vertical structures under pulsating waves (ALLSOP et al., 1995 and FRANCO et al., 1994) give lower overtopping than steep smooth slopes, but more than a rough rubble mound slope. The impulsive conditions give a different trend. First of all, the influence of the relative water depth is fairly small as all curves with different H_{m0}/h_s are quite close. For low vertical structures ($R_c/H_{m0} < 1.5$) there is hardly any difference between pulsating and impulsive conditions. The large difference is present for higher vertical structures and certainly for the very high structures. With impulsive conditions water is thrown high into the air, which means that overtopping occurs even for very high structures. The vertical distance that the discharge travels is more or less independent of the actual height of the structure. For $R_c/H_{m0} > 3$ the curves are almost horizontal.

Another way of comparing various structures is to show the influence of the slope angle on wave overtopping, and this has been done in Fig. 4.2. A vertical structure means $\cot\alpha = 0$. Steep smooth structures can roughly be described by $1 \leq \cot\alpha \leq 3$. Battered walls have $0 < \cot\alpha < 1$. Gentle slopes have roughly $\cot\alpha \geq 2$ or 3. Fig. 4.2 shows curves for two relative freeboards: $R_c/H_{m0} = 1.5$ & 3.0.

Of course similar conclusions can be drawn as for the previous comparison. Steep slopes give the largest overtopping, which reduces for gentler slopes; for a given wave condition and water level. Vertical slopes give less overtopping than steep smooth slopes, except for a high vertical structure under impulsive conditions.

Details of all equations used here are described in Chapter 5 (sloping smooth structures), Chapter 6 (rubble mound structures) and Chapter 7 (vertical structures).

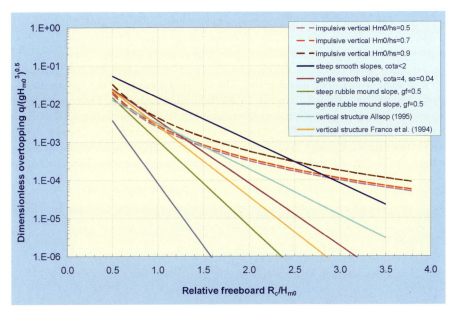

Fig. 4.1: Comparison of wave overtopping formulae for various kind of structures

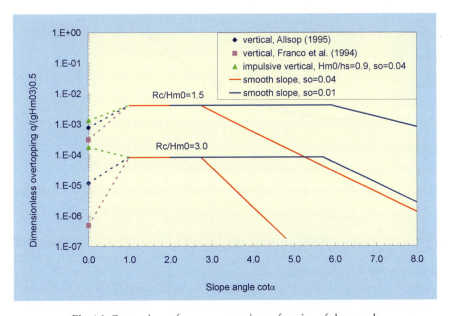

Fig. 4.2: Comparison of wave overtopping as function of slope angle

4.2.2 Overtopping volumes and V_{max}

Wave overtopping is a dynamic and irregular process and the mean overtopping discharge, q, does not cover this aspect. But by knowing the storm duration, t, and the number of overtopping waves in that period, N_{ow}, it is easy to describe this irregular and dynamic overtopping, if the overtopping discharge, q, is known. Each overtopping wave gives a certain overtopping volume of water, V and this can be given as a distribution

As many equations in this manual, the two-parameter Weibull distribution describes the behaviour quite well. This equation has a shape parameter, b, and a scale parameter, a. The shape parameter gives a lot of information on the type of distribution. Fig. 4.3 gives an overall view of some well-known distributions. The horizontal axis gives the probability of exceedance and has been plotted according to the Rayleigh distribution. The reason for this is that waves at deep water have a Rayleigh distribution and every parameter related to the deep water wave conditions, like shallow water waves or wave overtopping, directly show the deviation from such a Rayleigh distribution in the graph. A Rayleigh distribution should be a straight line in Fig. 4.3 and a deviation from a straight line means a deviation from the Rayleigh distribution.

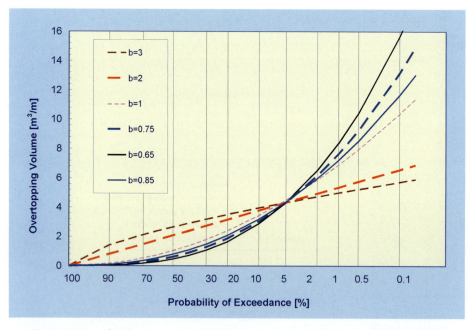

Fig. 4.3: Various distributions on a Rayleigh scale graph. A straight line (b = 2) is a Rayleigh distribution

When waves approach shallow water and the highest waves break, the wave distribution turns into a Weibull distribution with b > 2; also refer to Fig. 2.5. An example with b=3 is shown in Fig. 4.3 and this indicates that there are more large waves of similar height. The exponential distribution (often found for extreme wave climates) has b = 1 and shows that extremes become larger compared to most of the data. Such an exponential distribution would give a straight line in a log-linear graph.

The distribution of overtopping volumes for all kind of structures have average values even smaller than b = 1. Such a distribution is even steeper than an exponential distribution. It means that the wave overtopping process can be described by a lot of fairly small or limited overtopping volumes and a few very large volumes. The EA-manual (1999) gives various b-values (and according a-values), based on different and limited data sets. The b-values are mostly within the range 0.6 < b < 0.9. For comparison curves with b = 0.65 and 0.85 are given in Fig. 4.3. The curves are very similar, except that the extremes differ a little. It is for this reason that for smooth slopes an average b-value of 0.75 was chosen and not different values for various subsets of data. The same average value has been used for rubble mound structures, which makes smooth and rubble mound structures easy comparable. The exceedance probability, P_V, of an overtopping volume per wave is then similar to:

$$P_V = P\left(\underline{V} \leq V\right) = 1 - \exp\left[-\left(\frac{V}{a}\right)^{0.75}\right] \qquad 4.2$$

with:

$$a = 0.84 \cdot T_m \cdot \frac{q}{P_{ov}} = 0.84 \cdot T_m \cdot q \cdot N_w / N_{ow} = 0.84 \cdot q \cdot t / N_{ow} \qquad 4.3$$

Equation 4.3 shows that the scale parameter a, depends on the overtopping discharge, q, but also on the mean period, T_m, and probability of overtopping, N_{ow}/N_w, or which is similar, on the storm duration, t, and the actual number of overtopping waves N_w.

Equations for calculating the overtopping volume per wave for a given probability of exceedance, is given by Equation 4.2. The maximum overtopping during a certain event is fairly uncertain, as most maxima, but depends on the duration of the event. In a 6 hours period one may expect a larger maximum than only during 15 minutes. The maximum overtopping volume by only one wave during an event depends on the actual number of overtopping waves, N_{ow}, and can be calculated by:

$$V_{max} = a \cdot \left[\ln\left(N_{ow}\right)\right]^{4/3} \qquad 4.4$$

Chapters 5, 6 and 7 give formulae for smooth slopes, rubble mound slopes and vertical walls, respectively. In this Section and example is given between the mean overtopping discharge, q, and the maximum overtopping volume in the largest wave. Note that the mean overtopping is given in l/s per m width and that the maximum overtopping volume is given in l per m width.

As example a smooth slope with slope angle 1:4 is taken, a rubble mound slope with a steeper slope of 1:1.5 and a vertical wall. The storm duration has been assumed as 2 hours (the peak of the tide) and a fixed wave steepness of $s_{0m-1,0} = 0.04$ has been taken. Fig. 4.4 gives the $q - V_{max}$ lines for the three structures and for relatively small waves of $H_{m0} = 1$ m (red lines) and for fairly large waves of $H_{m0} = 2.5$ m (black lines).

A few conclusions can be drawn from Fig. 4.4. First of all, the ratio q/V_{max} is about 1000 for small q (roughly around 1 l/s per m) and about 100 for large q (roughly around 100 l/s per m). So, the maximum volume in the largest wave is about 100 – 1000 times larger than the mean overtopping discharge.

Secondly, the red lines are lower than the black lines, which means that for lower wave heights, *but similar mean discharge, q,* the maximum overtopping volume is also smaller. For example, a vertical structure with a mean discharge of 10 l/s per m gives a maximum volume of 1000 l per m for a 1 m wave height and a volume of 4000 l per m for a 2.5 m wave height.

Finally, the three different structures give different relationships, depending on the equations to calculate q and the equations to calculate the number of overtopping waves. More information can be found in Chapters 5, 6 and 7.

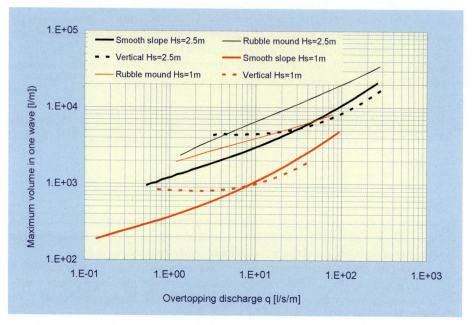

Fig. 4.4: Relationship between mean discharge and maximum overtopping volume in one wave for smooth, rubble mound and vertical structures for wave heights of 1 m and 2.5 m

4.2.3 Wave transmission by wave overtopping

Admissible overtopping depends on the consequences of this overtopping. If water is behind a structure, like for breakwaters and low-crested structures along the shore, large overtopping can be allowed as this overtopping will plunge into the water again. What happens is that the overtopping waves cause new waves behind the structure. This is called wave transmission and is defined by the wave transmission coefficient $K_t = H_{m0,t}/H_{m0,i}$, with $H_{m0,t}$ = transmitted significant wave height and $H_{m0,i}$ = incident significant wave height. The limits of wave transmission are $K_t = 0$ (no transmission) and 1 (no reduction in wave height). If a structure has its crest above water the transmission coefficient will never be larger than about 0.4–0.5.

Wave transmission has been investigated in the European DELOS project. For smooth sloping structures the following prediction formulae were derived:

$$K_t = \left[-0.3 \cdot \frac{R_c}{H_{m0,i}} + 0.75 \cdot \left(1 - \exp\left(-0.5 \cdot \xi_{0,p}\right)\right) \right] \cdot \left(\cos \beta\right)^{2/3} \qquad 4.5$$

with as a minimum $K_t = 0.075$ and maximum $K_t = 0.8$, and limitations $1 < \xi_{op} < 3$, $0° \le \beta \le 70$.

and $1 < B/H_i < 4$, and where β is the angle of wave attack and B is crest width (and not berm width).

Fig. 4.5 shows the transmission coefficient K_t as a function of the relative freeboard R_c/H_{m0} and for a smooth structure with slope angle $\cot\alpha = 4$ (a gentle smooth low-crested structure). Three wave steepnesses have been used: $s_{o,p} = 0.01$ (long waves), 0.03 and 0.05 (short wind waves). Also perpendicular wave attack has been assumed. Wave transmission decreases for increasing crest height and a longer wave gives more transmission. Wave overtopping can be calculated for the same structure and wave conditions, see Chapter 5 and Fig. 4.6. Also here a longer wave gives more wave overtopping.

The relationship between wave overtopping and transmission is found if both Fig.s are combined and Fig. 4.7 shows this relationship. For convenience the graphs are not made in a dimensionless way, but for a wave height of 3 m. A very small transmitted wave height of 0.1 m is only found if the wave overtopping is at least 30–50 l/s per m. In order to reach a transmitted wave height of about 1 m (one-third of the incident wave height) the wave overtopping should at least be 500–2500 l/s/m or 0.5–2.5 m³/s/m.

One may conclude that wave transmission is always associated with (very) large wave overtopping.

Wave transmission for rubble mound structures has also been investigated in the European DELOS project and the following prediction formulae were derived for wave transmission:

$$K_t = -0.4\, R_c/H_{m0} + 0.64\, B/H_{m0} - 0.31(1 - \exp(-0.5\xi_{op})) \quad \text{for } 0.075 \leq K_t \leq 0.8 \quad 4.6$$

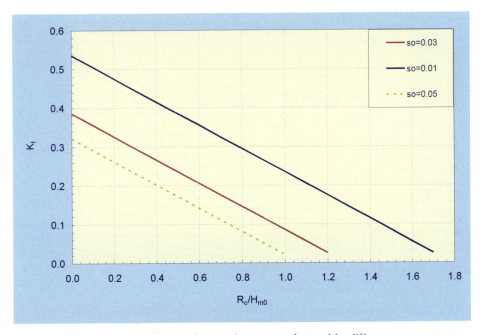

Fig. 4.5: Wave transmission for a gentle smooth structure of 1:4 and for different wave steepness

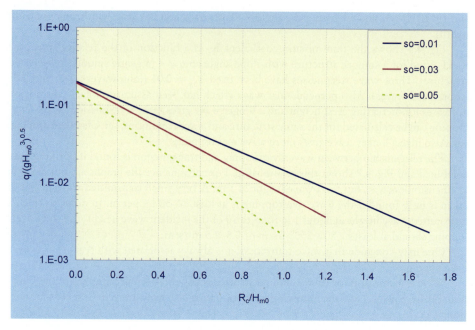

Fig. 4.6: Wave overtopping for a gentle smooth structure of 1:4 and for different wave steepness

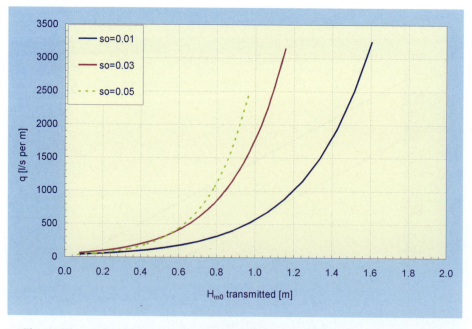

Fig. 4.7: Wave transmission versus wave overtopping for a smooth 1:4 slope and a wave height of H_{m0} = 3 m

Wave overtopping for a rubble mound structure with simple slope can be calculated by Equations in Chapter 6. A typical rubble mound structure has been used as example, with $\cot\alpha = 1.5$; 6–10 ton rock ($D_{n50} = 1.5$ m) as armour and a crest width of 4.5 m ($3\,D_{n50}$). A wave height of 3 m has been assumed with the following wave steepness: $s_{0m-1,0} = 0.01$ (long waves), 0.03 and 0.05 (short wind waves). In the calculations the crest height has been changed to calculate wave transmission as well as wave overtopping.

Fig. 4.8 gives the comparison. The graph shows that a longer wave ($s_{0m1,0} = 0.01$) gives more wave transmission, for the same overtopping discharge. The reason could be that wave overtopping is defined at the rear of the crest, where (without superstructure or capping wall), waves can penetrate through the armour layer at the crest and generate waves behind the structure. This is easier for longer waves.

In contrast to smooth structures, one may conclude that even without considerable wave overtopping discharge at the rear of the crest, there still might be considerable wave transmission through the structure. In this example transmitted wave heights between 0.5 m and 1 m are found for overtopping discharges smaller than 100–200 l/s per m. Only larger transmitted wave heights are associated with extreme large overtopping discharges of more than 500–1000 l/s per m.

A simple equation for wave transmission at vertical structures has been given by GODA (2000):

$$K_t = 0.45 - 0.3\, R_c/H_{m0} \qquad \text{for } 0 < R_c/H_{m0} < 1.25 \qquad 4.7$$

Wave overtopping for a simple vertical structure can be calculated by Equation 7.4. In both formulae only the relative crest height plays a role and no wave period, steepness or

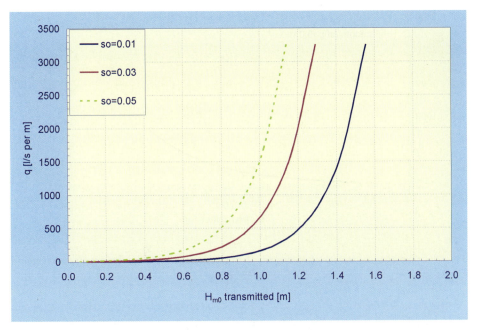

Fig. 4.8: Wave transmission versus wave overtopping discharge for a rubble mound structure, $\cot\alpha = 1.5$; 6–10 ton rock, $B = 4.5$ m and $H_{m0} = 3$ m

slope angle. A simple vertical structure has been used as example with a fixed incident wave height of H_{m0} = 3 m. Fig. 4.9 gives the comparison of wave overtopping and wave transmission, where in the calculations the crest height has been changed to calculate wave transmission as well as wave overtopping.

For comparison the same rubble mound structure has been used as the example in Fig. 4.8, with $\cot\alpha$ = 1.5; 6–10 ton rock (D_{n50} = 1.5 m) as armour, a crest width of 4.5 m (3 D_{n50}) and a wave steepness s_{0p} = 0.03. The curve for a smooth structure (Fig. 4.7) and for s_{0p} = 0.03 has been given too in Fig. 7.24.

A rubble mound structure gives more wave transmission than a smooth structure, under the condition that the overtopping discharge is similar. But a vertical structure gives even more transmission. The reason may be that overtopping water over the crest of a vertical breakwater always falls from a distance into the water, where at a sloping structure water flows over and/or through the structure.

One may conclude that even without considerable wave overtopping discharge at the crest of a vertical structure, there still might be considerable wave transmission. In this example of a vertical structure, transmitted wave heights between 0.5 m and 1 m are found for overtopping discharges smaller than 100–200 l/s per m.

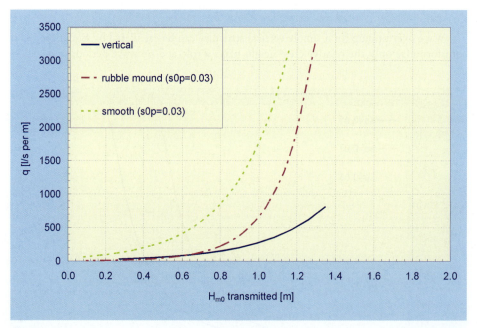

Fig. 4.9: Comparison of wave overtopping and transmission for a vertical, rubble mound and smooth structure

Fig. 4.10: Wave overtopping and transmission at breakwater IJmuiden, the Netherlands

4.3 PC-Overtopping

The programme PC-OVERTOPPING was made on the results of the Technical TAW Report "Wave run-up and wave overtopping at dikes" and is used for the 5-yearly safety assessment of all water defences in the Netherlands. The TAW Report has now in this Manual been replaced by Chapter 5 (dikes and embankments) and extended for rubble mound and vertical structures in Chapters 6 and 7. The programme was mainly based on a *dike type structure*. It means that the structure should be sloping, although a small vertical wall on top of the dike may be taken into account. Also roughness/permeability different from "smooth" can be taken into account, but not a crest with permeable and rough rock or armour units. In such a case the structure should be modelled up to the transition to the crest and other formulae should be used to take into account the effect of the crest (see Chapter 6).

The programme was set-up in such a way that almost *every sloping structure* can be modelled by an unlimited number of sections. Each section is given by x-y coordinates and each section can have its own roughness factor. The programme calculates almost all relevant overtopping parameters (except flow velocities and flow depths), such as:
- 2% run-up level;
- mean overtopping discharge;
- percentage of overtopping waves;
- overtopping volumes per wave (maximum and for every percentage defined by the user);
- required crest height for given mean overtopping discharges (defined by the user).

The main advantages of PC-OVERTOPPING are:
- Modelling of each sloping structure, including different roughness along the slope;
- Calculation of most overtopping parameters, not only the mean discharge.

The main disadvantage is:
- It does not calculate vertical structures and not a rough/permeable crest.

In order to show the capabilities of the programme an example will be given. Fig. 4.11 shows the cross-section of a dike with the design water level 1 m above CD. Different materials are used on the slope: rock, basalt, concrete asphalt, open concrete system and grass on the upper part of the structure. The structure has been schematised in Fig. 4.12 by x-y coordinates and a selection of the material of the top layer. The programme selects the right roughness factor.

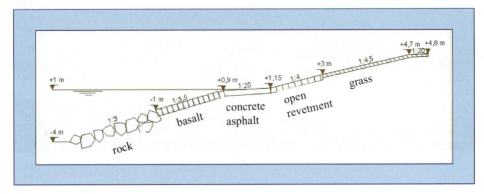

Fig. 4.11: Example cross-section of a dike

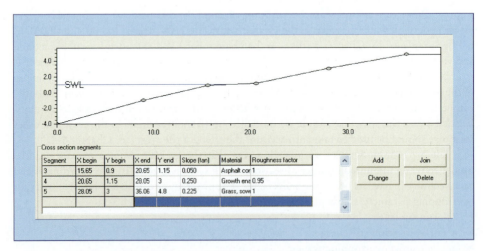

Fig. 4.12: Input of geometry by x-y coordinates and choice of top material

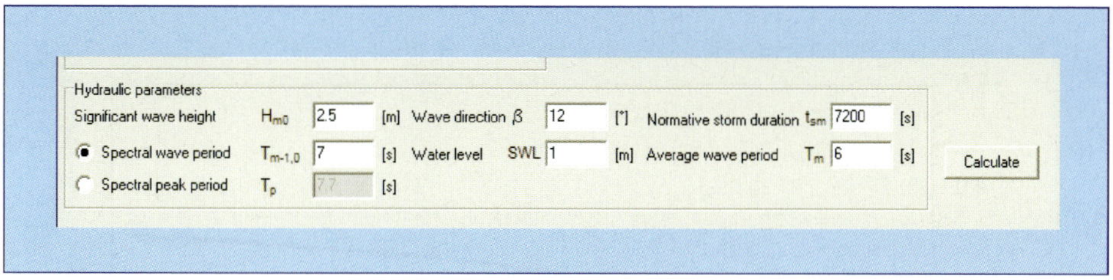

Fig. 4.13: Input file

The input parameters are the wave height, wave period (choice between the spectral parameter $T_{m-1,0}$ and the peak period T_p), the wave angle, water level (with respect to CD, the same level as used for the structure geometry) and finally the storm duration and mean period (for calculation of overtopping volumes, etc.). Fig. 4.13 gives the input file.

The output is given in three columns, see Fig. 4.14. The left column gives the 2% run-up level, the mean overtopping discharge and the percentage of overtopping waves. If the 2%-run-up level is higher than the actual dike crest, this level is calculated by extending the highest section in the cross-section. The middle column gives the required dike height for given mean overtopping discharges. Also here the highest section is extended, if required. Finally, in the right column the number of overtopping waves in the given storm duration are given, together with the maximum overtopping volume and other volumes, belonging to specified overtopping percentages (percentage of the number of overtopping waves).

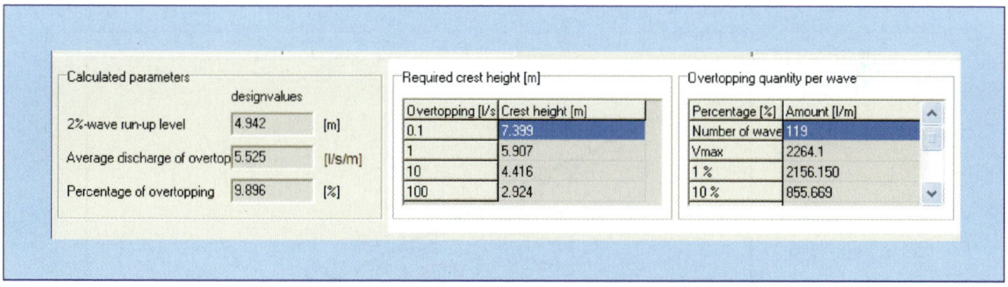

Fig. 4.14: Output of PC-OVERTOPPING

The programme also provides a kind of check whether found results of the 2% run-up level and mean overtopping discharge fall within measured ranges. All test results where the formulae were based on, are given in a run-up or overtopping graph, see Fig. 4.15 and Fig. 4.16. The graphs show the actual measured run-up or overtopping, including the effect of reductions due to roughness, berms, etc. The curve gives the maximum, which means a smooth straight slope with perpendicular wave attack. The programme then plots the calculated point in these graphs (the green point within the red circle).

54

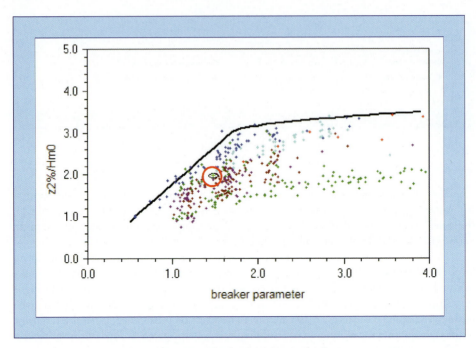

Fig. 4.15: Check on 2% run-up level

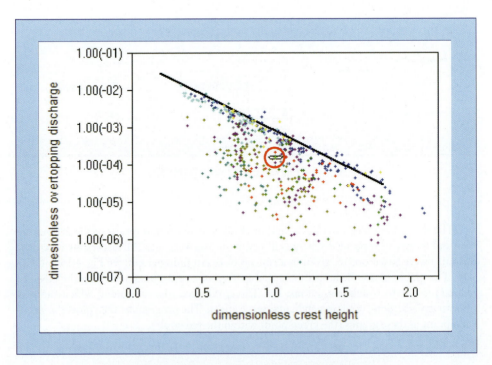

Fig. 4.16: Check on mean overtopping discharge

4.4 Neural network tools

Artificial neural networks fall in the field of artificial intelligence and can in this context be defined as systems that simulate intelligence by attempting to reproduce the structure of human brains. Neural networks are organised in the form of layers and within each layer there are one or more processing elements called 'neurons'. The first layer is the input layer and the number of neurons in this layer is equal to the number of input parameters. The last layer is the output layer and the number of neurons in this layer is equal to the number of output parameters to be predicted. The layers in between the input and output layers are the hidden layers and consist of a number of neurons to be defined in the configuration of the NN. Each neuron in each layer receives information from the preceding layer through the connections, carries out some standard operations and produces an output. Each connectivity has a weight factor assigned, as a result of the calibration of the neural network. The input of a neuron consists of a weighted sum of the outputs of the preceding layer; the output of a neuron is generated using a linear activation function. This procedure is followed for each neuron; the output neuron generates the final prediction of the neural network.

Artificial neural networks have applications in many fields and also in the field of coastal engineering for prediction of rock stability, forces on walls, wave transmission and wave overtopping. The development of an artificial neural network is useful if:
- the process to be described is complicated with a lot of parameters involved,
- there is a large amount of data.

Less complicated processes may be described by empirical formulae. This is also true for the process of wave overtopping, where many formulae exists, but always for a certain type of structure. Wave overtopping on all kind of structures can not be covered by only one formula, but a neural network is able to do this. A neural network needs a large amount of data to become useful for prediction. If the amount of data is too small, many predictions might be unreliable as the prediction will be out of range. But specially for the topic of wave overtopping there is an overwhelming amount of tests on all kinds of coastal structures and embankments.

This was the reason to start the European CLASH project. The result has been that two neural networks have been developed, one within CLASH and one along side of CLASH as a PhD-work. In both cases the neural network configuration was based on Fig. 4.17, where the input layer has 15 input parameters (β, h, H_{m0toe}, $T_{m-1,0toe}$, h_t, B_t, γ_f, $cot\alpha_d$, $cot\alpha_u$, R_c, B, h_b, $tan\ \alpha_b$, A_c, G_c) and 1 output parameter in the output layer (i.e. mean overtopping discharge, q). CLASH was focused on a three-layered neural network, where a configuration with one single hidden layer was chosen.

The development of an artificial neural network is a difficult task. All data should be checked thoroughly (rubbish in = rubbish out) and the training of a neural network needs special skills. The application of a developed neural network as a prediction tool, however, is easy and can be done by most practical engineers! It is for this reason that the CLASH neural network is part of this manual.

The application of the neural network is providing an Excel or ASCII input file with parameters, run the programme (push a button) and get a result file with mean overtopping discharge(s). Such an application is as easy as getting an answer from a formula programmed in Excel and does not need knowledge about neural networks. The advantages of the neural network are:
- it works for almost every structure configuration,
- it is easy to calculate trends instead of just one calculation with one answer.

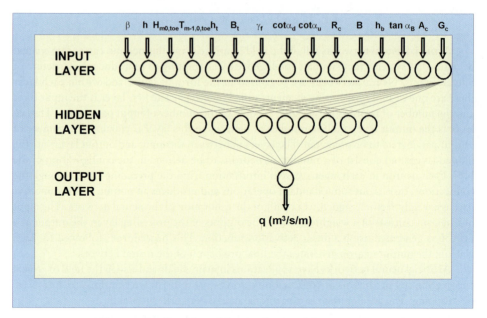

Fig. 4.17: Configuration of the neural network for wave overtopping

The input exists of 10 structural parameters and 4 hydraulic parameters. The hydraulic parameters are wave height, wave period, and wave angle and water depth just in front of the structure. The structural parameters describe almost every possible structure configuration by a toe (2 parameters), two structure slopes (including vertical and wave return walls), a berm (2 parameters) and a crest configuration (3 parameters). The tenth structural parameter is the roughness factor for the structure (γ_f) and describes the *average roughness of the whole structure*. Although guidance is given, estimation is not easy if the structure has different roughness on various parts of the structure. An overall view of possible structure configurations is shown in Fig. 4.18. It clearly shows that the neural network is able to cope with most structure types.

Very often one is not only interested in one calculation, but in more. As the input file has no limitations in number of rows (= number of calculations), it is easy to incrementally increase one or more parameters and to find a trend for a certain (design) measure. As example for calculation of a trend with the neural network tool an example cross-section of a rubble mound embankment with a wave wall has been chosen, see Fig. 4.19.

If, for example, the cross-section in Fig. 4.19 would have too much overtopping, the following measures could be considered:
- Increasing the crest
- Applying a berm
- Widening the crest
- Increasing only the crest wall.

Table 4.1 shows the input file with the first 6 calculations, where incremental increase of the crest will show the effect of raising the crest on the amount of wave overtopping. Calculations will give an output file with the mean overtopping discharge q (m^3/s per m width) and

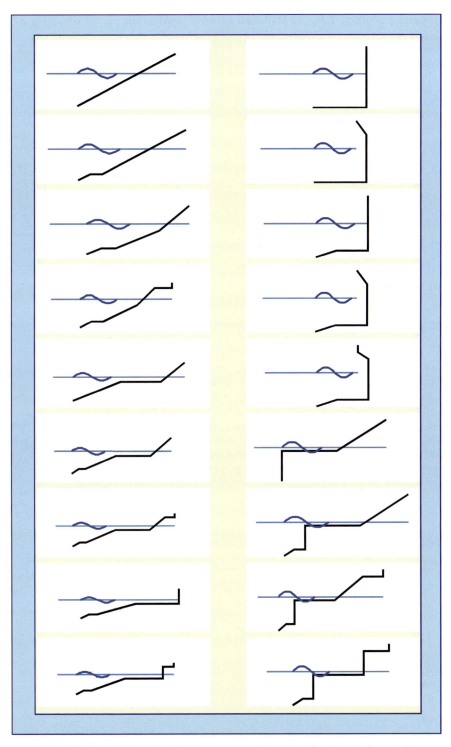

Fig. 4.18: Overall view of possible structure configurations for the neural network

with confidence limits. Table 4.2 shows an example which is the output belonging to the input in Table 4.1.

Table 4.1: Example input file for neural network with first 6 calculations

β	h	H_{m0}	$T_{m-1,0}$	h_t	B_t	γ_f	$\cot\alpha_d$	$\cot\alpha_u$	R_c	B	h_b	$\tan\alpha_B$	A_c	G_c
0	12	5	9.1	9	4	0.49	1.5	1.5	4	0	0	0	5	5
0	12	5	9.1	9	4	0.49	1.5	1.5	4.05	0	0	0	5.05	5
0	12	5	9.1	9	4	0.49	1.5	1.5	4.1	0	0	0	5.1	5
0	12	5	9.1	9	4	0.49	1.5	1.5	4.15	0	0	0	5.15	5
0	12	5	9.1	9	4	0.49	1.5	1.5	4.2	0	0	0	5.2	5
0	12	5	9.1	9	4	0.49	1.5	1.5	4.25	0	0	0	5.25	5

Table 4.2: Output file of neural network with confidence limits

q (m3/s/m)	2.50%	5.00%	25.00%	50.00%	75.00%	95.00%	97.50%	Nr.Pat.	REMARK I
5.90E-02	2.45E-02	2.77E-02	4.15E-02	5.91E-02	8.35E-02	0.1299	0.1591	1	4
5.64E-02	2.35E-02	2.64E-02	3.99E-02	5.58E-02	7.91E-02	0.1246	0.1516	2	4
5.40E-02	2.26E-02	2.49E-02	3.82E-02	5.33E-02	7.52E-02	0.119	0.1448	3	4
5.16E-02	2.19E-02	2.39E-02	3.69E-02	5.08E-02	7.17E-02	0.1133	0.1383	4	4
4.94E-02	2.07E-02	2.27E-02	3.55E-02	4.85E-02	6.89E-02	0.1079	0.1324	5	4
4.73E-02	1.99E-02	2.18E-02	3.38E-02	4.62E-02	6.60E-02	0.1033	0.1265	6	4

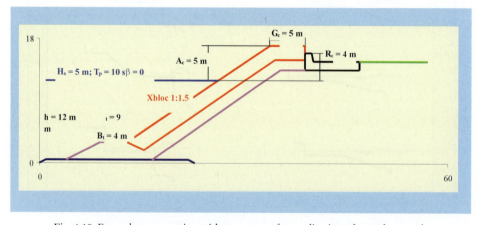

Fig. 4.19: Example cross-section with parameters for application of neural network

To make the input file for this example took 1 hour and resulted in 1400 rows or calculations. The calculation of the neural network took less than 10 seconds. The results were copied into the Excel input file and a resulting graph was made within Excel, which took another hour. Fig. 4.20 gives the final result, where the four trends are shown. The base situation had an overtopping discharge of 59 l/s per m. The graph clearly shows what measures are required to reduce the overtopping by for example a factor 10 (to 5.9 l/s per m) or to only 1 l/s per m. It also shows that increasing structure height is most effective, followed by increasing only the crest wall.

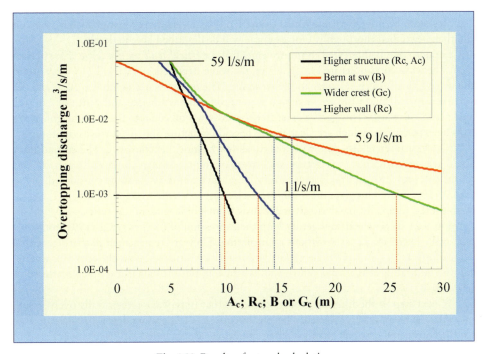

Fig. 4.20: Results of a trend calculation

At present two neural networks exist. One is the official neural network developed by Delft Hydraulics in the CLASH project. It runs as an executable and can be downloaded from the CLASH website or the Manual website. The other neural network has also been developed within CLASH, but as part of a PhD-thesis at Gent University (VERHAEGHE, 2005). The network was developed in MatLab® and actually an application can only be performed if the user has MatLab®, which is not often the case in the engineering world. An easier application has to be worked out: web based or executable.

The advantage of the Gent neural network is that it first decides whether there will be overtopping or not (classifier). If there is no overtopping it will give q = 0. If there is overtopping, it will quantify the overtopping with a similar network as the CLASH network (quantifier). This is certainly an advantage above the CLASH network. The CLASH network was only trained with overtopping data (tests with "no overtopping" were not considered) and, therefore, this network always gives a prediction of overtopping, also in the range where no overtopping should be expected.

4.5 Use of CLASH database

The EU-programme CLASH resulted in an extensive database with wave overtopping tests. Each test was described by 31 parameters as hydraulic and structural parameters, but also parameters describing the reliability and complexity of the test and structure. The database includes more than 10,000 tests and was set-up as an Excel database. The database, therefore, is nothing more than a matrix with 31 columns and more than 10,000 rows.

If a user has a specific structure, there is a possibility to look into the database and find more or less similar structures with measured overtopping discharges. It may even be possible that the structure has already been tested with the right wave conditions! Finding the right tests can be done by using filters in the Excel database. Every test of such a selection can then be studied thoroughly. One example will be described here in depth.

Suppose one is interested in improvement of a *vertical wall with a large wave return wall*. The wave conditions are $H_{m0\ toe}$ = 3 m, the wave steepness s_o = 0.04 ($T_{m-1,0}$ = 6.9 s) and the wave attack is perpendicular to the structure. The design water depth h = 10 m and the wave return wall starts 1 m above design water level and has a height and width of 2 m (the angle is 45° seaward). This gives a crest freeboard Rc = 3 m, equal to the wave height. Have tests been performed which are close to this specific structure and given wave conditions?

The first filter selects data with a vertical down slope, i.e. $\cot\alpha_d$ = 0. The second filter could select data with a wave return wall overhanging more than about 30° seaward. This means $\cot\alpha_u$ < −0.57. In first instance every large wave return wall can be considered, say at least 0.5 H_{m0} wide. This gives the third filter, selecting data with $-\cot\alpha_u * (A_c + h_b)/H_{m0} \geq 0.5$. With these 3 filters, the database gives 212 tests from 4 independent test series.

Fig. 4.21 shows the data together with the expression of FRANCO et al. (1994) for a vertical wall. There are 22 tests without overtopping. They are represented in the Fig. with a value of $q/(gH^3_{m0toe})^{1/2} = 10^{-7}$. The majority of the data is situated below the curve for a vertical wall, indicating that a wave return wall is efficient, but the data is too much scattered to be decisive.

A next step in the filtering process could be that only wave return walls overhanging more than 45° seaward are selected. This means $\cot\alpha_u$ < −1. The water depth is relatively large for the considered case and shallow water conditions are excluded if $h/H_{m0\ toe}$ > 3. Fig. 4.22 shows this further filtering process. The number of data has been reduced to 78 tests from only 2 independent series. In total 12 tests result in no overtopping. The data show the trend that the overtopping is in average about ten times smaller than for a vertical wall, given by the dashed line. But for $Rc/H_{m0\ toe}$ > 1 there are quite some tests without any overtopping.

As still quite some data are remaining in Fig. 4.22, it is possible to narrow the search area even further. With a wave steepness of s_o = 0.04 in the considered case, the wave steepness range can be limited to 0.03 < s_o < 0.05. The width of the wave return wall of 2 m gives with the wave height of 3 m a relative width of 0.67. The range can be limited to 0.5 < $-\cot\alpha_u * (A_c + h_b)/H_{m0}$ < 0.75. Finally, the transition from vertical to wave return wall is 1 m above design water level, giving $h_b/H_{m0\ toe}$ = −0.33. The range can be set at −0.5 < $h_b/H_{m0\ toe}$ < −0.2.

The final selection obtained after filtering is given in Fig. 4.23. Only 4 tests remain from one test series, one test resulted in no overtopping. The data give now a clear picture. For a relative freeboard lower than about $R_c/H_{m0\ toe}$ = 0.7 the overtopping will not be much different from the overtopping at a vertical wall. The wave return wall, however, becomes very efficient for large freeboards and even gives no overtopping for $R_c/H_{m0\ toe}$ > 1.2. For the structure considered with $R_c/H_{m0\ toe}$ = 1 the wave overtopping will be 20–40 times less then for a vertical wall and will probably amount to q = 0.5–2 l/s per m width. In this particular

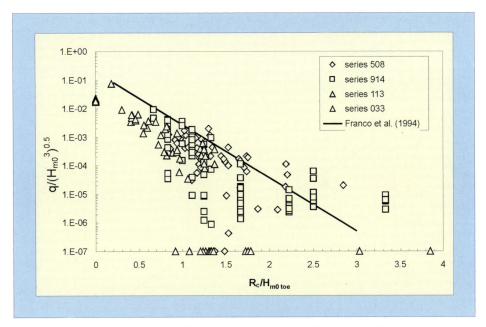

Fig. 4.21: Overtopping for large wave return walls; first selection

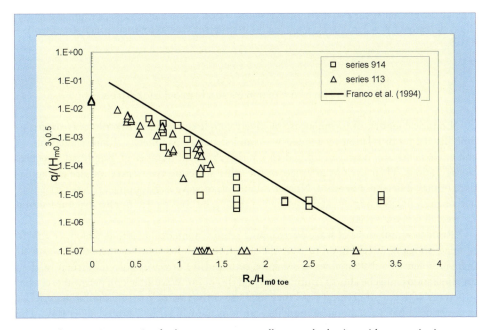

Fig. 4.22: Overtopping for large wave return walls; second selection with more criteria

case it was possible to find 4 tests in the database with very close similarities to the considered structure and wave conditions.

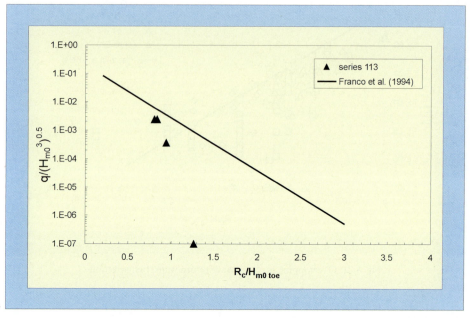

Fig. 4.23: Overtopping for a wave return wall with $s_o = 0.04$, seaward angle of 45°, a width of 2 m and a crest height of $R_c = 3$ m. For $H_{m0\,toe} = 3$ m the overtopping can be estimated from $R_c/H_{m0\,toe} = 1$

4.6 Outline of numerical model types

Empirical models or formulae use relatively simple equations to describe wave overtopping discharges in relation to defined wave and structure parameters. Empirical equations and coefficients are, however, limited to a relatively small number of simplified structure configurations. Their use out of range, or for other structure types, may require extrapolation, or may indeed not be valid. Numerical models of wave overtopping are less restrictive, in that any validated numerical model can; in theory; be configured for any structure within the overall range covered.

Realistic simulations of wave overtopping require numerical methods which are able to simulate shoaling, breaking on or over the structure, and possible overturning of waves. If there is violent or substantial wave breaking, or impulsive of waves onto the structure, then the simulations must be able to continue beyond this point. Wave attack on permeable coastal structures with a high permeability, such as those consisting of coarse granular material or large artificial blocks, cannot be modelled without modelling the porous media flow. The energy dissipation inside the permeable parts, the infiltration and seepage in the swash and backwash area, and the interactive flow between the external wave motion and the internal wave motion often cause the wave attack to be quite different from the flow on impermeable structures.

All of the processes described above occur during overtopping at structures, and all affect how the wave overtops and determine the peak and mean discharges. Additionally, physical model tests suggest that a sea state represented by 1000 random waves will give reasonably consistent results, but that shorter tests may show significant variations in extreme statistics. Any numerical model should therefore be capable of running similar numbers of waves.

There are no numerical models capable of meeting all of the above criteria accurately in a computationally effective or economical way, and it may be many more years before advances in computer technology allow these types of models to be used. There are, however, different model types each capable of meeting some of these criteria. They essentially fall into two principal categories: the nonlinear shallow water equation models (NLSW); and those based on the Navier-Stokes equations. Each of these generic types will now be discussed, with the emphasis on the range of applicability rather than the underlying mathematical principals.

4.6.1 Navier-Stokes models

The fluid motion for models based on the Navier-Stokes equations will generally be controlled by one of two principal techniques: the Volume of Fluid (VOF) method first described by HIRT & NICHOLS (1981); and the Smooth Particle Hydrodynamics (SPH) method as discussed by MONAGHAN (1994). Each of these models requires a detailed computational grid to be defined throughout the fluid domain, with solutions to the complex set of equations required at each grid-point before the simulation can continue. Restricted to only two dimensions, and for computational domains of only two or three wavelengths, these model types will typically take several minutes of computational time to simulate small fractions of a second of real time. In general SPH models take longer to run than VOF models.

An example of a model based on the Navier-Stokes equations is the VOF model SKYLLA. Developed to provide a wide range of applicability, high accuracy and a detailed description of the flow field for a wide range of structures, including permeable structures. It includes combined modelling of free surface wave motion, and porous media flow, and allows for simulations with large variations in the vertical direction in both the flow field and in the cross section of the structure. The internal wave motion is simulated within the porous media flow, and is valid for 2d incompressible flow with constant fluid mass density through a homogeneous isotropic porous medium. It is, nevertheless, restricted to regular waves, since irregular waves cannot be computed within manageable computational times.

Although computationally very expensive, these model types can provide descriptions of pressure and velocity fields within porous structures, and impulsive and breaking wave loads. Computation of wave transmission and wave run-up of monochromatic waves is possible, but the study of more than a few irregular overtopping waves is not yet possible.

4.6.2 Nonlinear shallow water equation models

The one-dimensional shallow water equations were originally developed for near horizontal, free-surface channel flows. The equations describe water depth and horizontal velocity in time and space, where vertical velocity is neglected, and only hydrostatic pressure is considered. The resulting nonlinear shallow water (NLSW) equations; derived from the

Navier-Stokes equations; simplify the mathematical problem considerably, allowing realistic, but simplified, real-time simulations to be computed.

The general restriction of these models is that they must be in shallow water (h/L < 0.05) if the model assumptions are to be preserved, and that waves entering into the computational domain have or will break. The fundamental mathematical assumption for NLSW models is that the waves travel as bores as described by HIBBERD & PEREGRINE (1979). At the crest of a sea defence structure, these models are able to continue computing as the flows either side of the crest separate, overtop or return.

ODIFLOCS (VAN GENT, 1994) is a one-dimensional, time-domain model which simulates the wave attack of perpendicular incident waves on permeable and impermeable coastal structures. The NLSW model is coupled to an internal porous media flow model (KOBAYASHI et al., 1987) that allows homogenous permeable structures to be modelled. This allows the modelling of infiltration and seepage phenomena, and the internal phreatic surface can be followed separately from the free surface flow. ODIFLOCS was developed to estimate permeability coefficients, wave transmission, magnitude of internal set-up, and the influence of spectral shape on wave run-up and overtopping.

The ANEMONE model suite developed by DODD (1998), comes as both a 1d and a 2d plan model, and also incorporates a porous media flow model for examining beaches (CLARKE et al., 2004). The landward boundaries, both for the free surface flow and for the internal boundary of the porous media flow, can be modelled as open or closed (non-reflecting or fully reflecting respectively). The model is capable of simulating storms of a 1000 waves or more at little computational cost, recording wave-by-wave and mean overtopping discharges.

These models, and others like them, are invaluable tools to examine the difference in overtopping performance when modifications to a scheme design are to be investigated. Long wave runs for a variety of sea states, for say a range of crest levels, is a problem well suited to these models. The overtopping discharges computed by these models should not, however, be relied upon as this is generally a function of how the model is set up for a given study: e.g. specification of the position of the seaward boundary in the model will affect the overtopping rate. The absolute difference in overtopping between two similar runs will usually produce reliable information.

4.7 Physical modelling

Physical model tests are an established and reliable method for determining mean wave overtopping discharges for arbitrary coastal structural geometries; additional levels of sophistication allow individual overtopping volumes to be measured. Typically at Froudian scales of 1:5 to 1:50, models represent the prototype structure in 2d or 3d, and frequently occurring and extreme storm events can be modelled. Wave flumes are usually of 0.3 to 1.5 m width with a depth of 0.5 to 1.0 m and fitted with a piston based wave paddle. Some form of wave absorbing system to compensate for waves reflected from the model structure is essential for overtopping studies in wave flumes. Wave basin models vary in size and complexity, and overtopping may often be measured at several locations on the model.

Physical model tests are particularly useful when assessing wave overtopping, as overtopping is affected by several factors whose individual and combined influences are still largely unknown and difficult to predict. The most common hydraulic parameters which influence wave overtopping are the significant wave height, the wave period, the wave direction (obliquity), and the water depth at the structure toe. The structural parameters are the slope, the berm width and level, the crest width and level, and the geometry of any crest/parapet wall. Where rock or concrete armour are used the porosity, permeability and placement pattern of armour units affect overtopping as does the roughness of the individual structural elements.

Due to the large number of relevant parameters, and the very complex fluid motion at the structure, theoretical approaches to wave overtopping are not well developed. Physical model tests, such as wave flume studies, are therefore commonly used to develop empirical formulae for predicting wave overtopping. These formulae do not assess wave overtopping discharges and individual volumes accurately, especially for low overtopping volumes, rather they provide an order of magnitude approximation. This is partially caused by so far unknown scale and model effects and the fact that only very limited field data exists. These scale and model effects are briefly discussed in the following section.

There are many cases where there are no reliable empirical overtopping prediction methods for a given structure geometry, or where the performance of a particular scheme to reduce overtopping is especially sensitive: e.g. where public safety is a concern. Alternatively, it may be that the consequences of overtopping are important: e.g. where overtopping waves cause secondary waves to be generated in the lee of the structure. For cases such as these, physical model testing may be the only reliable option for assessing overtopping.

4.8 Model and Scale effects

This section deals with model and scale effects resulting from scaled hydraulic models on wave overtopping. First, definitions will be given what scale effects and model effects are. Secondly, a methodology based on the current knowledge is introduced on how to account for these effects.

4.8.1 Scale effects

Scale effects result from incorrect reproduction of a prototype water-structure interaction in the scale model. Reliable results can only be expected by fulfilling Froude's and Reynolds' law simultaneously. This is however not possible so that scale effects cannot be avoided when performing scaled model tests.

Since gravity, pressure and inertial forces are the relevant forces for wave motion most models are scaled according to Froude's law. Viscosity forces are governed by Reynolds' law, elasticity by Cauchy's law and surface tension forces by Weber's law, and these forces have to be neglected for most models. All effects and errors resulting from ignoring these forces are called scale effects. The problem of the quantification of these scale effects is still unresolved.

4.8.2 Model and measurement effects

Model or laboratory effects originate from the incorrect reproduction of the prototype structure, geometry and waves and currents, or due to the boundary conditions of a wave flume (side walls, wave paddle, etc.). Modelling techniques have developed significantly, but there are still influences of model effects on hydraulic model results to be expected.

Measurement effects result from different measurement equipment used for sampling the data in the prototype and model situations. These effects, which are referred to as "measurement effects" may significantly influence the comparison of results between prototype and model, or two identical models. It is therefore essential to quantify the effects and the uncertainty related to the different techniques available.

4.8.3 Methodology

Following the aforementioned definitions the reasons for differences in between model and prototype data will sometimes be very difficult to assign to either model or scale effects. During CLASH, the major contributions to model effects were found to be wind since this is ignored in the hydraulic model. Despite the lack of wind, additional differences were found and assigned to be due to model effects. The following phenomena may give indications of the contributions of the most important model effects in addition to wind. The repeatability of tests showed that the wave parameters (H_{m0}, T_p, $T_{m-1,0}$) have a coefficient of variation of CoV~3 %, and for wave overtopping the differences between two wave flumes were CoV~13 % and CoV~10 %. Different time windows for wave analysis and different types of wave generation methods had no influence on the estimated wave parameters (CoV~3 %). The number of waves in the flume shows influence on wave overtopping, where a comparison of 200 compared to 1000 generated waves show differences in mean overtopping rates up to a value of 20 %. The position of the overtopping tray at the side of the flume showed also differences in overtopping rates (CoV~20 %) from results where the tray was located at the centre of the crest. This could be because of the different arrangement of the armour units in front of the overtopping tray or due to the influence of the side walls of the flume. More details on measurements and model effects are provided by KORTENHAUS et al. (2004a).

Scale effects have been investigated by various authors, and this has led to some generic rules that should be observed for physical model studies. Generally, water depths in the model should be much larger than h = 2.0 cm, wave periods larger than T = 0.35 s and wave heights larger than H_s = 5.0 cm to avoid the effects of surface tension; for rubble mound breakwaters the Reynolds number for the stability of the armour layer should exceed Re = 3×10^4; for overtopping of coastal dikes Re > 1×10^3; and the stone size in the core of rubble mound breakwaters has to be scaled according to the velocities in the core rather than the stone dimensions, especially for small models. The method for how this can be achieved is given in BURCHARTH et al. (1999). Furthermore, critical limits for the influence of viscosity and surface tension are given in Table 4.3, more details can be found in SCHÜTTRUMPF and OUMERACI (2005).

Table 4.3: Scale effects and critical limits

Process	Relevant forces	Similitude law	Critical limits
Wave propagation	Gravity force Friction forces Surface tension	Fr_W, Re_W, We	$Re_W > Re_{W,crit} = 1 \times 10^4$ $T > 0{,}35$ s; $h > 2{,}0$ cm
Wave breaking	Gravity force Friction forces Surface tension	Fr_W, Re_W, We	$Re_W > Re_{W,crit} = 1 \times 10^4$ $T > 0{,}35$ s; $h > 2{,}0$ cm
Wave run-up	Gravity force Friction forces Surface tension	Fr_A, Fr_q Re_q, We	$Re_q > Re_{q,crit} = 10^3$ $We > We_{crit} = 10$
Wave overtopping	Gravity force Friction forces Surface tension	Fr_A, Fr_q, Re_q, We	$Re_q > Re_{q,crit} = 10^3$ $We > We_{crit} = 10$

With: $Fr_W = c/(g \cdot h)^{1/2}$; $Fr_A = v_A/(g \cdot h_A)^{1/2}$; $Fr_q = v_A/(2 \cdot g \cdot R_u)$; $Re_W = c \cdot h/v$; $Re_q = (R_u - R_C)^2/(v \cdot T)$; $We = v_A \cdot h_A \cdot \rho_W/\sigma_W$

From observations in prototype and scaled models, a methodology was derived to account for those differences without specifically defining which model and measurement effects contribute how much. These recommendations are given in subsections 5.7 for dikes, 6.3.6 for rubble slopes, and 7.3.6 and 7.3.7 for vertical walls, respectively.

4.9 Uncertainties in predictions

Sections 4.2 to 4.4 have proposed various models to predict wave overtopping of coastal structures. These models will now be discussed with regard to their uncertainties.

4.9.1 Empirical Models

It has been discussed in section 1.5.4 that the model uncertainty concept uses a mean factor of 1.0 and a Gaussian distribution around the mean prediction. The standard deviation is derived from the comparison of model data and the model prediction.

This has two implications for design: Probabilistic design values for all empirical models used in this manual describe the mean approach for all underlying data points. This means that, for normally distributed variables, about 50 % of the data points exceed the prediction by the model, and 50 % are below the predicted values. This value should be used if probabilistic design methods are used.

The deterministic design value for all models will be given as the mean value plus one standard deviation, which in general gives a safer approach, and takes into account that model uncertainty for wave overtopping is always significant.

4.9.2 Neural Network

When running the Neural Network model the user will be provided with wave overtopping ratios based on the CLASH database and the Neural Network prediction (Section 4.4). Together with these results the user will also obtain the uncertainties of the prediction through the 5 % and 95 % confidence intervals.

Assuming a normal distribution of the results will allow an estimate of the standard deviation of the overtopping ratio and hence the whole Gaussian distribution. Results from the Neural Network prediction can then be converted to the methodology referenced in Section 4.8 by providing all other confidence intervals and exceedance probabilities required there. Details will be given when test cases will be investigated.

4.9.3 CLASH database

The CLASH database is described in Section 4.5. It provides a large dataset of available model data on wave overtopping of coastal structures. It should be mentioned that the model and scale effects approach introduced in Section 4.8 has not been applied to the database. Whenever these data are used for prototype predictions the user will have to check whether any scaling correction procedure is needed.

With respect to uncertainties all model results will contain variations in the measured overtopping ratios. Most of these variations will result from measurement and model effects as discussed earlier. Since the database is no real model but an additional source of data information no model uncertainty can be applied.

4.10 Guidance on use of methods

This manual is accompanied by an overall Calculation Tool outlined in Appendix A. This tool includes the elements:
- **Empirical Calculator** programmed with the main empirical overtopping equations in Chapters 5, 6 and 7 (limited to those that can be described explicitly, that is without iteration).
- **PC-Overtopping**, which codes all the prediction methods presented in Chapter 5 for mean overtopping discharge for (generally shallow sloped) sea dikes, see section 4.3.
- **Neural Network** tool developed in the CLASH research project to calculate mean overtopping for many types of structures, see section 4.4.
- **CLASH database**, a listing of input parameters and mean overtopping discharge from each of approximately 10,000 physical model tests on both idealised (research) test structures, and site specific designs. These data can be sifted to identify test results that may apply for configurations close to the reader's, see section 4.5.

None of these methods give the universally 'best' results. The most reliable method to be used will depend on the type and complexity of the structures, and the closeness with which it conforms to simplifying assumptions used in previous model testing (on which all of the methods above are inherently based).

In selecting which method to use, or which set of results to prefer when using more than one method, the user will need to take account of the origins of each method. It may also be important in some circumstances to use an alternative method to give a check on a particular

set of calculations. To assist these judgements, a set of simple rules of thumb are given here, but as ever, these should not be treated as universal truths.
- For **simple vertical, composite, or battered walls** which conform closely to the idealisations in Chapter 7, the results of the Empirical Calculator are likely to be more reliable than the other methods as test data for these structure types do not feature strongly in the Database or Neural Network, and PC-Overtopping is not applicable.
- For **simple sloped dikes** with a single roughness, many test data have been used to develop the formulae in the Empirical Calculator, so this may be the most reliable, and simplest to use / check. For dikes with multiple slopes or roughness, PC-Overtopping is likely to be the most reliable, and easiest to use, although independent checking may be more complicated. The Database or Neural Network methods may become more reliable where the structure starts to include further elements.
- For **armoured slopes and mounds**, open mound structures that most closely conform to the simplifying models may best be described by the formulae in the Empirical Calculator. Structures of lower permeability may be modelled using PC-Overtopping. Mounds and slopes with crown walls may be best represented by application of the Database or Neural Network methods.
- For **unusual or complex structures with multiple elements**, mean overtopping discharge may be most reliably predicted by PC-Overtopping (if applicable) or by the Database or Neural Network methods.
- For structures that require use of the Neural Network method, it is possible that the use of many data for other configurations to develop a single Neural Network method may introduce some averaging. It may therefore be appropriate to check in the Database to see whether there are already test data close to the configuration being considered. This procedure may require some familiarity with manipulating these types of test data.

In almost all instances, the use of any of these methods will involve some degree of simplification of the true situation. The further that the structure or design (analysis) conditions depart from the idealised configurations tested to generate the methods/tools discussed, the wider will be the uncertainties. Where the importance is high of the assets being defended, and/or the uncertainties in using these methods are large, then the design solution may require use of site specific physical model tests, as discussed in section 4.6.

5. Coastal dikes and embankment seawalls

5.1 Introduction

An exact mathematical description of the wave run-up and wave overtopping process for coastal dikes or embankment seawalls is not possible due to the stochastic nature of wave breaking and wave run-up and the various factors influencing the wave run-up and wave overtopping process. Therefore, wave run-up and wave overtopping for coastal dikes and embankment seawalls are mainly determined by empirical formulas derived from experimental investigations. The influence of roughness elements, wave walls, berms, etc. is taken into account by introducing influence factors. Thus, the following chapter is structured as follows.

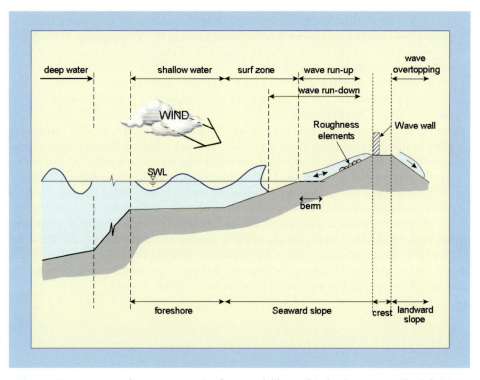

Fig. 5.1: Wave run-up and wave overtopping for coastal dikes and embankment seawalls: definition sketch. See Section 1.4 for definitions

First, wave run-up will be described as a function of the wave breaking process on the seaward slope for simple smooth and straight slopes. Then, wave overtopping is discussed with respect to average overtopping discharges and individual overtopping volumes. The influencing factors on wave run-up and wave overtopping like berms, roughness elements, wave walls and oblique wave attack are handled in the following section. Finally, the overtopping flow depth and the overtopping flow velocities are discussed as the direct influencing parameters to the surface of the structure. The main calculation procedure for coastal dikes and embankment seawalls is given in Fig. 5.2.

Simple slopes and conditions	Determinstic design	Probabilistic design
Wave run-up height	Eq. 5.4	Eq. 5.3
Mean overtopping discharge	Eq. 5.9	Eq. 5.8
Mean overtopping discharge (shallow foreshore)	Eq. 5.10	Eq. 5.11
Individual overtopping volumes	Eq. 5.35	Eq. 5.35
Complex slopes and conditions		
Effect of surface roughness	Table 5.2	Table 5.2
Effect of oblique waves	Eq. 5.23 for wave run-up Eq. 5.24 for wave overtopping	Eq. 5.23 for wave run-up Eq. 5.24 for wave overtopping
Effect of composite slopes	Eq. 5.26	Eq. 5.26
Effect of berms	Eq. 5.27	Eq. 5.27
Effect of wave wall	Eq. 5.34	Eq. 5.34
Overtopping Flow Parameters	**Flow depth**	**Flow velocity**
Seaward slope	5.41	5.43
Dike crest	5.44	5.45
Landward slope	5.44	5.49

Fig. 5.2: Main calculation procedure for coastal dikes and embankment seawalls

Definitions of, and detailed descriptions of, wave run-up, wave overtopping, foreshore, structure, slope, berm and crest height are given in Section 1.4 and are not repeated here.

5.2 Wave run-up

The wave run-up height is defined as the vertical difference between the highest point of wave run-up and the still water level (SWL) (Fig. 5.3). Due to the stochastic nature of the incoming waves, each wave will give a different run-up level. In the Netherlands as well as in Germany many dike heights have been designed to a wave run-up height $R_{u2\%}$. This is the wave run-up height which is exceeded by 2 % of the number of incoming waves at the toe of the structure. The idea behind this was that if only 2 % of the waves reach the crest of a dike or embankment during design conditions, the crest and inner slope do not need specific protection measures other than clay with grass. It is for this reason that much research in the past has been focused on the 2%-wave run-up height. In the past decade the design or safety assessment has been changed to allowable overtopping instead of wave run-up. Still a good prediction of wave run-up is valuable as it is the basic input for calculation of number of overtopping waves over a dike, which is required to calculate overtopping volumes, overtopping velocities and flow depths.

The general formula that can be applied for the 2%-wave run-up height is given by Equation 5.1: The relative wave run-up height $R_{u,2\%}/H_{m0}$ in Equation 5.11 is related to the breaker parameter $\xi_{m-1,0}$. The breaker parameter or surf similarity parameter $\xi_{m-1,0}$ relates the slope steepness $\tan \alpha$ (or 1/n) to the wave steepness $s_{m-1,0} = H_{m0}/L_0$ and is often used to distinguish different breaker types, see Section 1.4.

$$\frac{R_{u2\%}}{H_{m0}} = c_1 \cdot \gamma_b \cdot \gamma_f \cdot \gamma_\beta \cdot \xi_{m-1,0} \text{ with a maximum of } \frac{R_{u2\%}}{H_{m0}} = \gamma_f \cdot \gamma_\beta \left(c_2 - \frac{c_3}{\sqrt{\xi_{m-1,0}}} \right)$$

where:

$R_{u2\%}$ = wave run-up height exceeded by 2% of the incoming waves [m]

c_1, c_2 and c_3 = empirical coefficients [-] with   5.1
γ_b = influence factor for a berm [-]
γ_f = influence factor for roughness elements on a slope [-]
γ_β = influence factor for oblique wave attack [-]
$\xi_{m-1,0}$ = breaker parameter = $\tan\alpha/(s_{m-1,0})^{0.5}$ [-]
ξ_{tr} = transition breaker parameter between breaking and non-breaking waves (refer to Section 1.4.3)

The relative wave run-up height increases linearly with increasing $\xi_{m-1,0}$ in the range of breaking waves and small breaker parameters less than ξ_{tr}. For non-breaking waves and higher breaker parameter than ξ_{tr} the increase is less steep as shown in Fig. 5.4 and becomes more or less horizontal. The relative wave run-up height $R_{u,2\%}/H_{m0}$ is also influenced by: the geometry of the coastal dike or embankment seawall; the effect of wind; and the properties of the incoming waves.

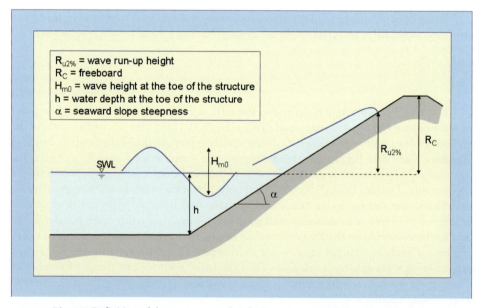

Fig. 5.3: Definition of the wave run-up height $R_{u2\%}$ on a smooth impermeable slope

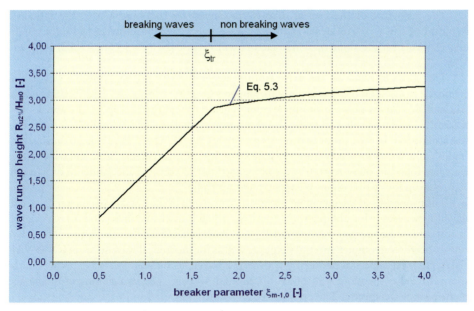

Fig. 5.4: Relative Wave run-up height $R_{u2\%}/H_{m0}$ as a function of the breaker parameter $\xi_{m-1,0}$, for smooth straight slopes

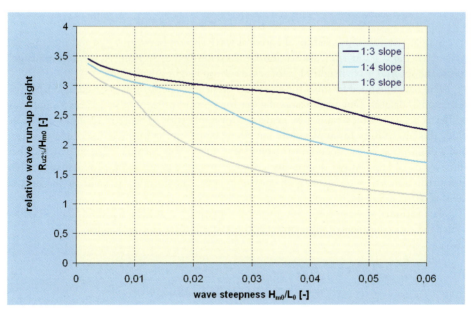

Fig. 5.5: Relative Wave run-up height $R_{u2\%}/H_{m0}$ as a function of the wave steepness for smooth straight slopes

The geometry of the coastal dike is considered by the slope tan α, the influence factor for a berm γ_b, the influence factor for a wave wall γ_v and the influence factor for roughness elements on the slope γ_f. These factors will be discussed in Sections 5.3.2, 5.3.4 and 5.3.5.

The effect of wind on the wave run-up-height for smooth impermeable slopes will mainly be focused on the thin layer in the upper part of the run-up. As described in Section 1.4, very thin layers of wave run-up are not considered and the run-up height was defined where the run-up layer becomes less than 1–2 cm. Wind will not have a lot of effect then. This was also proven in the European programme OPTICREST, where wave run-up on an actual smooth dike was compared with small scale laboratory measurements. Scale and wind effects were not found in those tests. It is recommended not to consider the influence of wind on wave run-up for coastal dikes or embankment seawalls.

The properties of the incoming waves are considered in the breaker parameter $\xi_{m-1,0}$ and the influence factor for oblique wave attack γ_β which is discussed in Section 5.3.3. As given in Section 1.4, the spectral wave period $T_{m-1,0}$ is most suitable for the calculation of the wave run-up height for complex spectral shapes as well as for theoretical wave spectra (JONSWAP, TMA, etc.). This spectral period $T_{m-1,0}$ gives more weight to the longer wave periods in the spectrum and is therefore well suited for all kind of wave spectra including bi-modal and multi-peak wave spectra. The peak period T_p, which was used in former investigations, is difficult to apply in the case of bi-modal spectra and should not be applied for multi peak or flat wave spectra as this may lead to large inaccuracies. Nevertheless, the peak period T_p is still in use for single peak wave spectra and there is a clear relationship between the spectral period $T_{m-1,0}$ and the peak period T_p for traditional single peak wave spectra:

$$T_p = 1.1\, T_{m-1,0} \qquad 5.2$$

Similar relationships exist for theoretical wave spectra between $T_{m-1,0}$ and other period parameters like T_m and $T_{m0,1}$, see Section 1.4. As described in Section 1.4, it is recommended to use the spectral wave height H_{m0} for wave run-up height calculations.

The recommended formula for wave run-up height calculations is based on a large (international) dataset. Due to the large dataset for all kind of sloping structures a significant scatter is present, which cannot be neglected for application. There are several ways to include this uncertainty for application, but all are based on the formula describing the mean and a description of the uncertainty around this mean. This formula is given first and then three kinds of application: deterministic design or safety assessment; probabilistic design; and prediction or comparison with measurements. The formula is valid in the area of $0.5 < \gamma_b \cdot \xi_{m-1,0} \leq 8$ to 10.

The formula of wave run-up is given by Equation 5.3 and by the solid line in Fig. 5.6 which indicates the average value of the 2 % measured wave run-up heights.

$$\frac{R_{u2\%}}{H_{m0}} = 1.65 \cdot \gamma_b \cdot \gamma_f \cdot \gamma_\beta \cdot \xi_{m-1,0}$$

$$\text{with a maximum of } \frac{R_{u2\%}}{H_{m0}} = 1.00 \cdot \gamma_b \cdot \gamma_f \cdot \gamma_\beta \left(4.0 - \frac{1.5}{\sqrt{\xi_{m-1,0}}} \right) \qquad 5.3$$

Fig. 5.4 shows the influence of the wave steepness for different slopes on the dimensionless wave run-up height $R_{u2\%}/H_{m0}$.

The wave run-up formulas are given in Fig. 5.6 together with measured data from small and large scale model tests. All data were measured under perpendicular wave attack and in relatively deep water at the dike toe without any significant wave breaking in front of the dike toe.

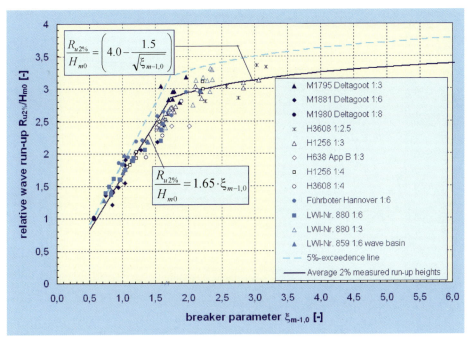

Fig. 5.6: Wave run-up for smooth and straight slopes

The statistical distribution around the average wave run-up height is described by a normal distribution with a variation coefficient $\sigma' = \sigma / \mu = 0.07$. It is this uncertainty which should be included in application of the formula. Exceedance lines, for example, can be drawn by using $R_{u2\%} / H_{m0} = \mu \pm x \cdot \sigma = \mu \pm x \cdot \sigma' \cdot \mu$, where μ is the prediction by Equation 5.3, $\sigma = \sigma' \cdot \mu$ the standard deviation, and x a factor of exceedance percentage according to the normal distribution. For example x = 1.64 for the 5 % exceedance limits and x = 1.96 for the 2.5 % exceedance limits. The 5 % upper exceedance limit is also given in Fig. 5.6.

$$\frac{R_{u2\%}}{H_{m0}} = 1.75 \cdot \gamma_b \cdot \gamma_f \cdot \gamma_\beta \cdot \xi_{m-1,0} \qquad \text{with a maximum of}$$

$$\frac{R_{u2\%}}{H_{m0}} = 1.00 \cdot \gamma_f \cdot \gamma_\beta \left(4.3 - \frac{1.6}{\sqrt{\xi_{m-1,0}}}\right) \qquad 5.4$$

Deterministic design or safety assessment: For design or a safety assessment of the crest height, it is advised not to follow the average trend, but to include the uncertainty of the prediction. In many international standards and guidelines a safety margin of about one standard deviation is used in formulae where the formula itself has significant scatter. Note that this standard deviation does not take into account the uncertainty of the parameters used, like the wave height and period. The equation for deterministic calculations is given by the dashed line in Fig. 5.7 together with the equation for probabilistic design. Equation 5.4 is recommended for deterministic calculations.

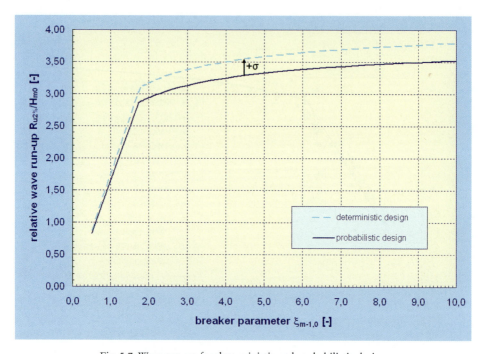

Fig. 5.7: Wave run-up for deterministic and probabilistic design

Probabilistic design: Besides deterministic calculations, probabilistic calculations can be made to include the effect of uncertainties of all parameters or to find optimum levels including the wind, wave and surge statistics. For probabilistic calculations Equation 5.3 is used together with the normal distribution and variation coefficient of $\sigma' = 0.07$.

Prediction or comparison of measurements: The wave run-up equation can also be used to predict a measurement in a laboratory (or in real situations) or to compare with measurements performed. In that case Equation 5.3 for the average wave run-up height should be used, preferably with for instance the 5 % upper and lower exceedance lines.

The influence factors γ_b, γ_f and γ_β where derived from experimental investigations. A combination of influence factors is often required in practice which reduces wave run-up and wave overtopping significantly. Systematic investigations on the combined influence of wave obliquity and berms showed that both influence factors can be used independently without any interactions. Nevertheless, a systematic combination over the range of all influence factors and all combinations was not possible until now. Therefore, further research is recommended if the overall influence factor $\gamma_b \gamma_f \gamma_\beta$ becomes lower than 0.4.

5.2.1 History of the 2% value for wave run-up

The choice for 2% has been made long ago and was probably arbitrary. The first international paper on wave run-up, mentioning the 2 % wave run-up, is ASBECK et al., 1953. The formula $R_{u2\%} = 8\ H_{m0}\ \tan\alpha$ has been mentioned there (for 5 % wave steepness and gentle smooth slopes, and this formula has been used for the design of dikes till 1980. But the choice for the 2 % was already made there.

The origin stems from the closing of the Southern Sea in the Netherlands in 1932 by the construction of a 32 km long dike (Afsluitdijk). This created the fresh water lake IJsselmeer and in the 45 years after closure about half of the lake was reclaimed as new land, called polders. The dikes for the first reclamation (North East Polder) had to be designed in 1936/1937. It is for this reason that in 1935 en 1936 a new wind-wave flume was built at Delft Hydraulics and first tests on wave run-up were performed in 1936. The final report on measurements (report M101), however, was issued in 1941 "due to lack of time". But the measurements had been analysed in 1936 to such a degree that "the dimensions of the dikes of the North East Polder could be established". That report could not be retrieved from Delft Hydraulics' library. The M101 report gives only the 2 % wave run-up value and this must have been the time that this value would be the right one to design the crest height of dikes.

Further tests from 1939–1941 on wave run-up, published in report M151 in 1941, however, used only the 1% wave run-up value. Other and later tests (M422, 1953; M500, 1956 and M544, 1957) report the 2% value, but for completeness give also the 1 %, 10 %, 20 % and 50 %.

It can be concluded that the choice for the 2% value was made in 1936, but the reason why is not clear as the design report itself could not be retrieved.

5.3 Wave overtopping discharges

5.3.1 Simple slopes

Wave overtopping occurs if the crest level of the dike or embankment seawall is lower than the highest wave run-up level R_{max}. In that case, the freeboard RC defined as the vertical difference between the still water level (SWL) and the crest height becomes important (Fig. 5.3). Wave overtopping depends on the freeboard RC and increases for decreasing freeboard height RC. Usually wave overtopping for dikes or coastal embankments is described by an average wave overtopping discharge q, which is given in m³/s per m width, or in litres/s per m width.

An average overtopping discharge q can only be calculated for quasi-stationary wave and water level conditions. If the amount of water overtopping a structure during a storm is required, the average overtopping discharge has to be calculated for each more or less constant storm water level and constant wave conditions.

Many model studies were performed to investigate the average overtopping discharge for specific dike geometries or wave conditions. For practical purposes, empirical formulae were fitted through experimental model data which obey often one of the following expressions:

$$Q_* = Q_0 (1 - R_*)^b \quad \text{or} \quad Q_* = Q_0 \exp(-b \cdot R_*) \qquad 5.5$$

Q_* is a dimensionless overtopping discharge, R_* is a dimensionless freeboard height, Q_0 describes wave overtopping for zero freeboard and b is a coefficient which describes the specific behaviour of wave overtopping for a certain structure. SCHÜTTRUMPF (2001) summarised expressions for the dimensionless overtopping discharge Q_* and the dimensionless freeboard height R_*.

As mentioned before, the average wave overtopping discharge q depends on the ratio between the freeboard height R_C and the wave run-up height R_u:

$$\frac{R_C}{R_u} \qquad 5.6$$

The wave run-up height R_u can be written in a similar expression as the wave run-up height $R_{u,2\%}$ giving the following relative freeboard height:

$$\frac{R_C}{c_{u,1} \cdot \xi_{m-1,0} \cdot H_{m0} \cdot \gamma_b \cdot \gamma_f \cdot \gamma_\beta \cdot \gamma_v} \quad \text{for breaking waves and a maximum of}$$

$$\frac{R_C}{c_{u,2} \cdot H_{m0} \cdot \gamma_f \cdot \gamma_\beta} \quad \text{for non-breaking waves} \qquad 5.7$$

The relative freeboard does not depend on the breaker parameter $\xi_{m-1,0}$ for non breaking waves (Fig. 5.8), as the line is horizontal.

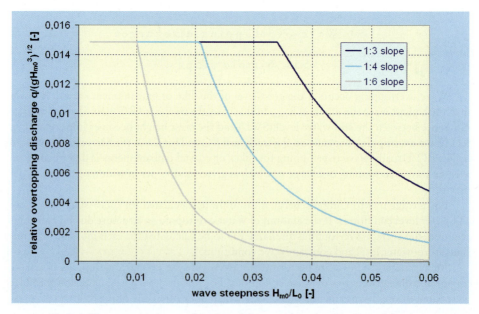

Fig. 5.8: Wave overtopping as a function of the wave steepness H_{m0}/L_0 and the slope

The dimensionless overtopping discharge $Q^* = q/(gH^3_{m0})^{1/2}$ is a function of the wave height, originally derived from the weir formula.

Probabilistic design and prediction or comparison of measurements ($\xi_{m-1,0}<5$): TAW (2002) used these dimensionless factors to derive the following overtopping formulae for breaking and non-breaking waves, which describe the *average* overtopping discharge:

$$\frac{q}{\sqrt{g \cdot H_{m0}^3}} = \frac{0.067}{\sqrt{\tan\alpha}} \gamma_b \cdot \xi_{m-1,0} \cdot \exp\left(-4.75 \frac{R_C}{\xi_{m-1,0} \cdot H_{m0} \cdot \gamma_b \cdot \gamma_f \cdot \gamma_\beta \cdot \gamma_v}\right) \quad 5.8$$

with a maximum of: $\dfrac{q}{\sqrt{g \cdot H_{m0}^3}} = 0.2 \cdot \exp\left(-2.6 \dfrac{R_C}{H_{m0} \cdot \gamma_f \cdot \gamma_\beta}\right)$

The reliability of Equation 5.8 is described by taking the coefficients 4.75 and 2.6 as normally distributed stochastic parameters with means of 4.75 and 2.6 and standard deviations σ = 0.5 and 0.35 respectively. For probabilistic calculations Equation 5.8 should be taken together with these stochastic coefficients. For predictions of measurements or comparison with measurements also Equation 5.8 should be taken with, for instance, 5 % upper and lower exceedance curves.

Equation 5.8 is given in Fig. 5.9 together with measured data for breaking waves from different model tests in small and large scale as well as in wave flumes and wave basins. In addition, the 5 % lower and upper confidence limits are plotted.

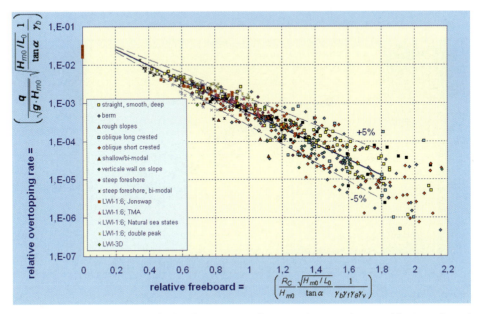

Fig. 5.9: Wave overtopping data for breaking waves and overtopping Equation 5.8 with 5 % under and upper exceedance limits

Data for non-breaking waves are presented in Fig. 5.10 together with measured data, the overtopping formula for non-breaking waves and the 5% lower and upper confidence limits.

Equation 5.8 gives the averages of the measured data and can be used for probabilistic calculations or predictions and comparisons with measurements.

Deterministic design or safety assessment ($\xi_{m-1,0}<5$): For deterministic calculations in design or safety assessment it is strongly recommended to increase the average discharge by about one standard deviation. Thus, Equation 5.9 should be used for deterministic calculations in design and safety assessment:

$$\frac{q}{\sqrt{g \cdot H_{m0}^3}} = \frac{0.067}{\sqrt{\tan\alpha}} \gamma_b \cdot \xi_{m-1,0} \cdot \exp\left(-4.3 \frac{R_C}{\xi_{m-1,0} \cdot H_{m0} \cdot \gamma_b \cdot \gamma_f \cdot \gamma_\beta \cdot \gamma_v}\right) \quad 5.9$$

with a maximum of: $\dfrac{q}{\sqrt{g \cdot H_{m0}^3}} = 0.2 \cdot \exp\left(-2.3 \dfrac{R_C}{H_{m0} \cdot \gamma_f \cdot \gamma_\beta}\right)$

A comparison of the two recommended formulas for deterministic design and safety assessment (Equation 5.8) and probabilistic calculations (Equation 5.9) for breaking and non-breaking waves is given in Fig. 5.11 and Fig. 5.12.

In the case of very heavy breaking on a shallow foreshore the wave spectrum is often transformed in a flat spectrum with no significant peak. In that case, long waves are present and influencing the breaker parameter $\xi_{m-1,0}$. Other wave overtopping formulas (equation 5.10 and 5.11) are recommended for shallow and very shallow foreshores to avoid a large

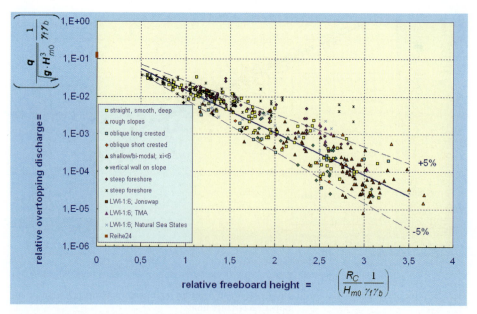

Fig. 5.10: Wave overtopping data for non-breaking waves and overtopping Equation 5.9 with 5% under and upper exceedance limits

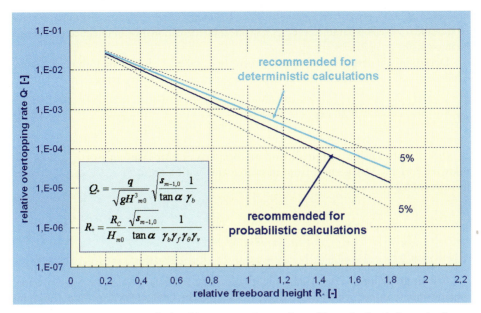

Fig. 5.11: Wave overtopping for breaking waves – Comparison of formulae for design and safety assessment and probabilistic calculations

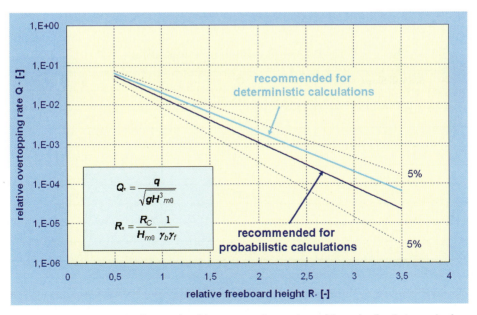

Fig. 5.12: Wave overtopping for non-breaking waves – Comparison of formulae for design and safety assessment and probabilistic calculations

underestimation of wave overtopping by using formulas 5.8 and 5.9. Since formulas 5.8 and 5.9 are valid for breaker parameters $\xi_{m-1,0}<5$ a linear interpolation is recommended for breaker parameters $5<\xi_{m-1,0}<7$.

Deterministic design or safety assessment ($\xi_{m-1,0}>7$): The following formula is recommended including a safety margin for deterministic design and safety assessment.

$$\frac{q}{\sqrt{g \cdot H_{m0}^3}} = 0.21 \cdot \exp\left(-\frac{R_C}{\gamma_f \cdot \gamma_\beta \cdot H_{m0} \cdot (0.33 + 0.022 \cdot \xi_{m-1,0})}\right) \qquad 5.10$$

Probabilistic design and prediction or comparison of measurements ($\xi_{m-1,0}>7$): The following formula was derived from measurements with a mean of −0.92 and a standard deviation of 0.24:

$$\frac{q}{\sqrt{g \cdot H_{m0}^3}} = 10^c \cdot \exp\left(-\frac{R_C}{\gamma_f \cdot \gamma_\beta \cdot H_{m0} \cdot (0.33 + 0.022 \cdot \xi_{m-1,0})}\right) \qquad 5.11$$

British guidelines recommend a slightly different formula to calculate wave overtopping for smooth slopes, which was originally developed by OWEN (1980) for smooth sloping and bermed seawalls:

$$\frac{q}{T_m \cdot g \cdot H_S} = Q_0 \cdot \exp\left(-b \cdot \frac{R_C}{T_m \sqrt{g \cdot H_S}}\right) \qquad 5.12$$

where Q_0 and b are empirically derived coefficients given in Table 5.1 (for straight slopes only).

Table 5.1: Owen's coefficients for simple slopes

Seawall Slope	Q_0	b
1:1	7.94E-3	20.1
1:1.5	8.84E-3	19.9
1:2	9.39E-3	21.6
1:2.5	1.03E-2	24.5
1:3	1.09E-2	28.7
1:3.5	1.12E-2	34.1
1:4	1.16E-2	41.0
1:4.5	1.20E-2	47.7
1:5	1.31E-2	55.6

Equation 5.2 uses the mean period T_m instead of the spectral wave period $T_{m-1,0}$ and has therefore the limitation of normal single peaked spectra which are not too wide or too narrow. Furthermore H_s, being $H_{1/3}$, was used and not H_{m0}, although this only makes a difference in shallow water. Equation 5.12 looks quite different to 5.8 and 5.9, but actually can be rewritten to a shape close to the breaking wave part of these formulae:

$$\frac{q}{\sqrt{g \cdot H_s^3}} = Q_0 / \sqrt{s_{0,m}} \cdot \exp\left(-b \frac{R_C}{H_s \sqrt{s_{0,m}}}\right) \qquad 5.13$$

If now tanα would be introduced in Equation 5.12 with a fit to the coefficients in Table 5.1, a similar formula as the breaking wave Equation 5.9 would be found. One restriction is that Equation 5.12 has no maximum for breaking waves, which may lead to significant over predictions for steep slopes and long waves.

The original data of OWEN (1980) were also used to develop Equations 5.8 and 5.9, which avoids the interpolation effort of the Owen formula for different slope angles given in Table 5.1 and overcomes other restrictions described above. But there is no reason not to use Equation 5.12 within the limits of application.

Zero Freeboard: Wave overtopping for zero freeboard (Fig. 5.13) becomes important if a dike or embankment seawall is overtopping resistant (for example a low dike of asphalt) and the water level comes close to the crest. SCHÜTTRUMPF (2001) performed model tests for different straight and smooth slopes in between 1:3 and 1:6 to investigate wave overtopping for zero freeboard and derived the following formula (σ' = 0.14), which should be used for probabilistic design and prediction and comparison of measurements (Fig. 5.14):

$$\frac{q}{\sqrt{g \cdot H_{m0}^3}} = 0.0537 \cdot \xi_{m-1,0} \qquad \text{for: } \xi_{m-1,0} < 2.0$$

$$\frac{q}{\sqrt{g \cdot H_{m0}^3}} = \left(0.136 - \frac{0.226}{\xi_{m-1,0}^3}\right) \qquad \text{for: } \xi_{m-1,0} \geq 2.0 \qquad 5.14$$

For deterministic design or safety assessment it is recommended to increase the average overtopping discharge in Equation 5.14 by about one standard deviation.

Negative freeboard: If the water level is higher than the crest of the dike or embankment seawall, large overtopping quantities overflow/overtop the structure. In this situation, the amount of water flowing to the landward side of the structure is composed by a part which can be attributed to overflow ($q_{overflow}$) and a part which can be attributed to overtopping ($q_{overtop}$). The part of overflowing water can be calculated by the well known weir formula for a broad crested structure:

$$q_{overflow} = 0.6 \cdot \sqrt{g \cdot \left|-R_C^3\right|} \qquad 5.15$$

where R_C is the (negative) relative crest height and $-R_c$ is the overflow depth [m]

The effect of wave overtopping ($q_{overtop}$) is accounted for by the overtopping discharge at zero freeboard ($R_C = 0$) in Equation 5.14 as a first guess.

The effect of combined wave run-up and wave overtopping is given by the superposition of overflow and wave overtopping as a rough approximation:

$$q = q_{overflow} + q_{overtop} = 0.6 \cdot \sqrt{g \cdot \left|-R_C^3\right|} + 0.0537 \cdot \xi_{m-1,0} \cdot \sqrt{g \cdot H_{m0}^3} \qquad 5.16$$

$$\text{for: } \xi_{m-1,0} < 2.0$$

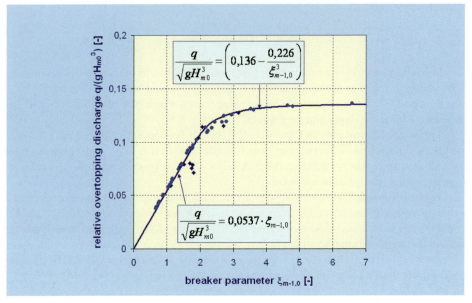

Fig. 5.13: Dimensionless overtopping discharge for zero freeboard (SCHÜTTRUMPF, 2001)

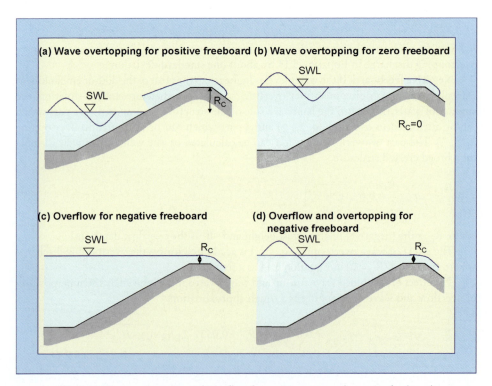

Fig. 5.14: Wave overtopping and overflow for positive, zero and negative freeboard

Wave overtopping is getting less important for increasing overflow depth R_C. An experimental verification of Equation 5.16 is still missing. Therefore, no distinction was made here for probabilistic and deterministic design.

5.3.2 Effect of roughness

Most of the seadikes and embankment seawalls are on the seaward side covered either by grass (Fig. 5.15), by asphalt (Fig. 5.16) or by concrete or natural block revetment systems (Fig. 5.17). Therefore, these types of surface roughness (described as smooth slopes) were often used as reference in hydraulic model investigations and the influence factor for surface roughness γ_f of these smooth slopes for wave heights greater than about 0.75 m is equal to $\gamma_f = 1.0$.

Fig. 5.15: Dike covered by grass (photo: SCHÜTTRUMPF)

Fig. 5.16: Dike covered by asphalt (photo: SCHÜTTRUMPF)

Fig. 5.17: Dike covered by natural bloc revetment (photo: SCHÜTTRUMPF)

For significant wave heights H_s less than 0.75 m, grass influences the run-up process and lower influence factors γ_f are recommended by TAW (1997) (Fig. 5.18). This is due to the relatively greater hydraulic roughness of the grass surface for thin wave run-up depths.

$$\gamma_f = 1.15\, H_s^{0.5} \text{ for grass and } H_s < 0.75 \text{ m} \qquad 5.17$$

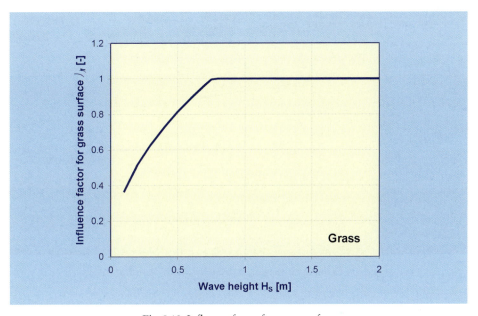

Fig. 5.18: Influence factor for grass surface

Roughness elements (Fig. 5.19) or slopes partly covered by rock are often used to increase the surface roughness and to reduce the wave run-up height and the wave overtopping rate. Roughness elements are either used to influence the wave run-up or the wave run-down process. Fig. 5.21 shows the influence of artificial roughness elements on the wave run-up and run-down process. Roughness elements are applied either across the entire slope or for parts of the slope which should be considered during the calculation process.

Available data on the influence of surface roughness on wave run-up and wave overtopping are based on model tests in small, but mainly in large scale, in order to avoid scale effects. A summary of typical types of surface roughness is given in Table 5.2.

The influence factors for roughness elements apply for $\gamma_b \cdot \xi_{m-1,0} < 1.8$, increase linearly up to 1.0 for $\gamma_b \cdot \xi_{m-1,0} = 10$ and remain constant for greater values. The efficiency of artificial roughness elements such as blocks or ribs depends on the width of the block or rib f_b, the height of the blocks f_h and the distance between the ribs f_L. The optimal ratio between the height and the width of the blocks was found to be $f_h/f_b = 5$ to 8 and the optimal distance between ribs is $f_L/f_b = 7$. When the total surface is covered by blocks or ribs and when the height is at least $f_h/H_{m0} = 0.15$, then the following minimum influence factors are found:

Block, 1/25 of total surface covered $\gamma_{f,min} = 0.85$
Block, 1/9 of total surface covered $\gamma_{f,min} = 0.80$
Ribs, $f_L/f_b = 7$ apart (optimal) $\gamma_{f,min} = 0.75$

Fig. 5.19: Example for roughness elements (photo: SCHÜTTRUMPF)

Table 5.2: Surface roughness factors for typical elements

Reference type	γ_f
Concrete	1.0
Asphalt	1.0
Closed concrete blocks	1.0
Grass	1.0
Basalt	0.90
Small blocks over 1/25 of surface	0.85
Small blocks over 1/9 of surface	0.80
¼ of stone setting 10 cm higher	0.90
Ribs (optimum dimensions)	0.75

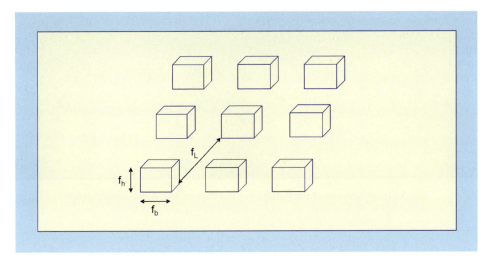

Fig. 5.20: Dimensions of roughness elements

A greater block or rib height than $f_h/H_{m0} = 0.15$ has no further reducing effect. If the height is less, then an interpolation is required:

$$\gamma_f = 1 - (1 - \gamma_{f,\,\text{min}}) \cdot \left(\frac{f_h}{0.15 \cdot H_{m0}} \right) \quad \text{for: } f_b/H_{m0} < 0.15 \qquad 5.18$$

As already mentioned, roughness elements are mostly applied for parts of the slope. Therefore, a reduction factor is required which takes only this part of the slope into account.

It can be shown that roughness elements have no or little effect below $0.25 \cdot R_{u2\%,\text{smooth}}$ below the still water line and above $0.50 \cdot R_{u2\%,\text{smooth}}$ above the still water line. The resulting influence factor γ_f is calculated by weighting the various influence factors $\gamma_{f,i}$ and by including the lengths L_i of the appropriate sections i in between SWL$-0.25 \cdot R_{u2\%\text{smooth}}$ and SWL $+ 0.50 \cdot R_{u2\%\text{smooth}}$:

$$\gamma_f = \frac{\sum_{i=1}^{n} \gamma_{f,i} \cdot L_i}{\sum_{i=1}^{n} L_i} \qquad 5.19$$

It appears that roughness elements applied only under water (with a smooth upper slope) have no effect and, in such a case, should be considered as a smooth slope. For construction purposes, it is recommended to restrict roughness elements to their area of influence. The construction costs will be less than covering the entire slope by roughness elements.

The effect of roughness elements on wave run-up may be reduced by debris between the elements.

Fig. 5.21: Performance of roughness elements showing the degree of turbulence

5.3.3 Effect of oblique waves

Wave run-up and wave overtopping can be assumed to be equally distributed along the longitudinal axis of a dike. If this axis is curved, wave run-up or wave overtopping will certainly increase for concave curves; with respect to the seaward face; due to the accumulation of wave run-up energy. Similarly, wave run-up and overtopping will decrease for convex curves, due to the distribution of wave run-up energy. No experimental investigations are known concerning the influence of a curved dike axis and the spatial distribution of wave run-up and wave overtopping yet.

Only limited research is available on the influence of oblique wave attack on wave run-up and wave overtopping due to the complexity and the high costs of model tests in wave basins. Most of the relevant research was performed on the influence of long crested waves and only few investigations are available on the influence of short crested waves on wave run-up and wave overtopping. Long crested waves have no directional distribution and wave crests are parallel and of infinite width. Only swell coming from the ocean can be regarded as a long crested wave. In nature, storm waves are short crested (Fig. 5.23). This means, that wave crests are not parallel, the direction of the individual waves is scattered around the main direction and the crests of the waves have a finite width. The directional spreading might be characterized by the directional spreading width σ or the spreading factor s. Relations between these parameters are approximately:

$$s = \frac{2-\sigma^2}{\sigma^2} \text{ or: } \sigma = \sqrt{\frac{2}{s+1}} \qquad 5.20$$

The directional spreading width is $\sigma = 0°$ ($s = \infty$) for long crested waves. Results of systematic research on the influence of oblique wave attack on wave run-up and wave overtopping under short crested wave conditions are summarized in EAK (2002) and TAW (2002). The data of this systematic research were summarized in Fig. 5.24. Data for long crested waves are not presented here.

The angle of wave attack β is defined at the toe of the structure after any transformation on the foreshore by refraction or diffraction as the angle between the direction of the waves and the perpendicular to the long axis of the dike or revetment as shown in Fig. 5.22. Thus, the direction of wave crests approaching parallel to the dike axis is defined as β = 0° (perpendicular wave attack). The influence of the wave direction on wave run-up or wave overtopping is defined by an influence factor γ_β:

$$\gamma_\beta = \frac{R_{u2\%;\beta>0°}}{R_{u2\%;\beta=0°}} \text{ for wave run-up} \qquad 5.21$$

$$\gamma_\beta = \frac{q_{\beta>0°}}{q_{\beta=0°}} \text{ for wave overtopping} \qquad 5.22$$

For practical purposes, it is recommended to use the following expressions for short crested waves to calculate the influence factor γ_β for wave run-up:

$$\begin{aligned} \gamma_\beta &= 1 - 0.0022|\beta| \text{ for} : 0° \leq \beta \leq 80° \\ \gamma_\beta &= 0.824 \text{ for} : |\beta| > 80° \end{aligned} \qquad 5.23$$

and wave overtopping

$$\begin{aligned} \gamma_\beta &= 1 - 0.0033|\beta| \text{ for} : 0° \leq \beta \leq 80° \\ \gamma_\beta &= 0.736 \text{ for} : |\beta| > 80° \end{aligned} \qquad 5.24$$

New model tests (SCHÜTTRUMPF et al. (2003)) indicate that formulae 5.21 and 5.22 overestimate slightly the reduction of wave run-up and wave overtopping for small angles of wave attack. The influence of wave direction on wave run-up or wave overtopping can be even neglected for wave directions less than $|\beta| = 20°$.

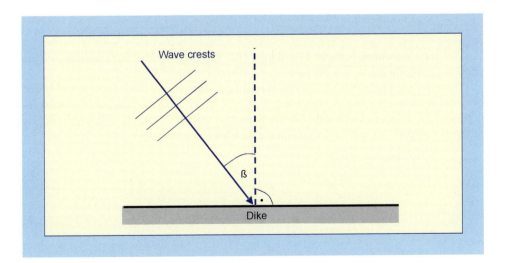

Fig. 5.22: Definition of angle of wave attack β

Fig. 5.23: Short crested waves resulting in wave run-up and wave overtopping (photo: Zitscher)

For wave directions $80° < |\beta| \leq 110°$ waves are diffracted around the structure and an adjustment of the wave height H_{m0} and the wave period $T_{m-1,0}$ are recommended:

H_{m0} is multiplied by $\dfrac{110-|\beta|}{30}$ $T_{m-1,0}$ is multiplied by $\sqrt{\dfrac{110-|\beta|}{30}}$

For wave directions between $110° < |\beta| \leq 180°$ wave run-up and wave overtopping are set to $R_{u2\%} = 0$ and $q = 0$.

No significant influence of different spreading widths s (s = ∞, 65, 15 and 6) was found in model tests. As long as some spreading is present, short-crested waves behave similar independent of the spreading width. The main point is that short-crested oblique waves give different wave run-up and wave overtopping than long-crested waves.

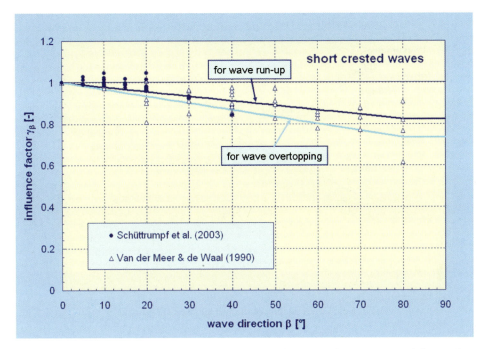

Fig. 5.24: Influence factor γ_β for oblique wave attack and short crested waves, measured data are for wave run-up

5.3.4 Composite slopes and berms

(a) Average slopes: Many dikes do not have a straight slope from the toe to the crest but consist of a composite profile with different slopes, a berm or multiple berms. A characteristic slope is required to be used in the breaker parameter $\xi_{m-1,0}$ for composite profiles or bermed profiles to calculate wave run-up or wave overtopping. Theoretically, the run-up process is influenced by a change of slope from the breaking point to the maximum wave run-up height. Therefore, often it has been recommended to calculate the characteristic slope from the point of wave breaking to the maximum wave run-up height. This approach needs some calculation effort, because of the iterative solution since the wave run-up height $R_{u2\%}$ is unknown. For the breaking limit a point on the slope can be chosen which is 1.5 H_{m0} below the still water line.

It is recommended to use also a point on the slope 1.5 H_{m0} above water as a first estimate to calculate the characteristic slope and to exclude a berm (Fig. 5.25).

$$1^{st} \text{ estimate: } \tan\alpha = \frac{3 \cdot H_{m0}}{L_{Slope} - B} \qquad 5.25$$

As a second estimate, the wave run-up height from the first estimate is used to calculate the average slope (L_{Slope} has to be adapted see Fig. 5.26):

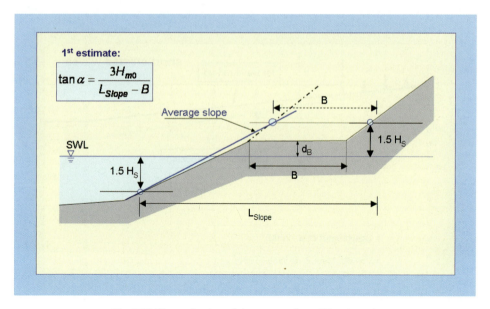

Fig. 5.25: Determination of the average slope (1st estimate)

$$2^{nd} \text{ estimate:} \quad \tan\alpha = \frac{\left(1.5 \cdot H_{m0} + R_{u2\%\,(from\ 1st\ estimate)}\right)}{L_{Slope} - B} \qquad 5.26$$

If the run-up height or 1.5 H_{m0} comes above the crest level, then the crest level must be taken as the characteristic point above SWL.

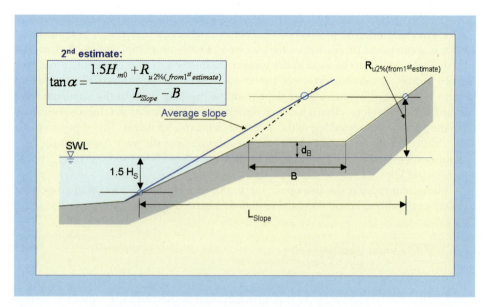

Fig. 5.26: Determination of the average slope (2nd estimate)

(b) Influence of Berms: A berm is a part of a dike profile in which the slope varies between horizontal and 1:15 (see Section 1.4 for a detailed definition). A berm is defined by the width of the berm B and by the vertical difference d_B between the middle of the berm and the still water level (Fig. 5.27). The width of the berm B may not be greater than $0.25 \cdot L_0$. If the berm is horizontal, the berm width B is calculated according to Fig. 5.27. The lower and the upper slope are extended to draw a horizontal berm without changing the berm height dB. The horizontal berm width is therefore shorter than the angled berm width. dB is zero if the berm lies on the still water line. The characteristic parameters of a berm are defined in Fig. 5.27.

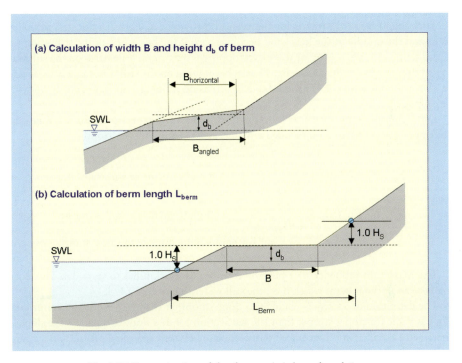

Fig. 5.27: Determination of the characteristic berm length L_{Berm}

Fig. 5.28: Typical berms (photo: SCHÜTTRUMPF)

A berm reduces wave run-up or wave overtopping. The influence factor γ_b for a berm consists of two parts.

$$\gamma_b = 1 - r_B(1 - r_{db}) \quad \text{for: } 0.6 \leq \gamma_b \leq 1.0 \qquad 5.27$$

The first part (r_B) stands for the width of the berm L_{Berm} and becomes zero if no berm is present.

$$r_B = \frac{B}{L_{Berm}} \qquad 5.28$$

The second part (r_{db}) stands for the vertical difference d_B between the still water level (SWL) and the middle of the berm and becomes zero if the berm lies on the still water line. The reduction of wave run-up or wave overtopping is maximum for a berm on the still water line and decreases with increasing d_B. Thus, a berm lying on the still water line is most effective. A berm lying below $2 \cdot H_{m0}$ or above $R_{u2\%}$ has no influence on wave run-up and wave overtopping.

Different expressions are used for r_{dB} in Europe. Here an expression using a cosine-function for r_{db} (Fig. 5.29) is recommended which is also used in PC-Overtopping.

$$r_{db} = 0.5 - 0.5 \cos\left(\pi \frac{d_b}{R_{u2\%}}\right) \quad \text{for a berm above still water line}$$

$$r_{db} = 0.5 - 0.5 \cos\left(\pi \frac{d_b}{2 \cdot H_{m0}}\right) \quad \text{for a berm below still water line} \qquad 5.29$$

$$r_{db} = 1 \text{ for berms lying outside the area of influence}$$

The maximum influence of a berm is actually always limited to $\gamma_B = 0.6$. This corresponds to an optimal berm width B on the still water line of $B = 0.4 \cdot L_{berm}$.

The definition of a berm is made for a slope smoother than 1:15 while the definition of a slope is made for slopes steeper than 1:8, see Section 1.4. If a slope or a part of the slope lies in between 1:8 and 1:15 it is required to interpolate between a bermed profile and a straight profile. For wave run-up this interpolation is written by:

$$R_{u2\%} = R_{u2\%(1:8slope)} + \left(R_{u2\%,(Berm)} - R_{u2\%,(1:8slope)}\right) \cdot \frac{(1/8 - \tan\alpha)}{(1/8 - 1/15)} \qquad 5.30$$

A similar interpolation procedure should be followed for wave overtopping.

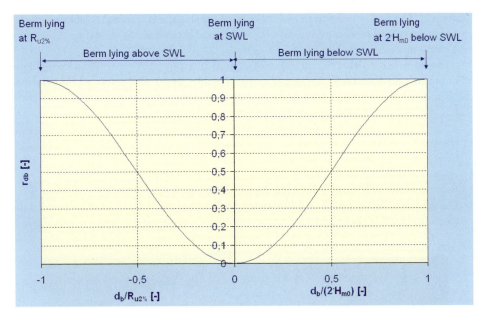

Fig. 5.29: Influence of the berm depth on factor r_{dh}

5.3.5 Effect of wave walls

In some cases a vertical or very steep wall is placed on the top of a slope to reduce wave overtopping. Vertical walls on top of the slope are often adopted if the available place for an extension of the basis of the structure is restricted. These are essentially relatively small walls and not large vertical structures such as caissons and quays (these are treated separately in

Fig. 5.30: Sea dike with vertical crest wall (photo: HOFSTEDE)

Chapter 7). The wall must form an essential part of the slope, and sometimes includes a berm or part of the crest. The effectiveness of a wave wall to reduce wave run-up and wave overtopping might be significant (Fig. 5.31).

The knowledge about the influence of vertical or steep walls on wave overtopping is quite limited and only a few model studies are available. Based on this limited information, the influence factors for a vertical or steep wall apply for the following studied application area:

- the average slope of 1.5 H_{m0} below the still water line to the foot of the wall (excluding a berm) must lie between 1:2.5 to 1:3.5.
- the width of all berms together must be no more than 3 H_{m0}.
- the foot of the wall must lie between about 1.2 H_{m0} under and above the still water line;
- the minimum height of the wall (for a high foot) is about 0.5 H_{m0}. The maximum height (for a low foot) is about 3 H_{m0}.

Fig. 5.31: Influence of a wave wall on wave overtopping (photo: SCHÜTTRUMPF)

It is possible that work will be performed to prepare guidance for wave overtopping for vertical constructions on a dike or embankment, in the future. Until then the influence factors below can be used within the application area described. Wave overtopping for a completely vertical walls is given in Chapter 7 of this manual.

For wave overtopping a breaker parameter is required, as for wave run-up. A vertical wall soon leads to a large value for the breaker parameter when determining an average slope as described in Fig. 5.25. This means that the waves will not break. The wall will be on top of the slope, possibly even above the still water line, and the waves will break on the slope before the wall. In order to maintain a relationship between the breaker parameter and the type of breaking on the dike slope, the steep or vertical wall must be drawn as a slope 1:1 when determining the average slope. This slope starts at the foot of the vertical wall. The average slope and the influence of any berm must be determined with a 1:1 slope instead of the actual steep slope or vertical wall, according to the procedure given before.

Furthermore, the overtopping for a vertical wall on the top of a dike is smaller than for a 1:1 slope on top of a dike profile. The influence factor for a vertical wall on a slope is $\gamma_V = 0.65$. For a 1:1 slope, this influence factor is $\gamma_V = 1$. Interpolation must be performed for a wall that is steeper than 1:1 but not vertical:

$$\gamma_v = 1.35 - 0.0078 \cdot \alpha_{wall} \qquad 5.31$$

where α_{wall} is the angle of the steep slope in degrees (between 45° for a 1:1 slope and 90° for a vertical wall).

The method to calculate the reduction factor for vertical walls is very limited to the given conditions. Therefore, it is recommended to use the Neural Network for more reliable calculations.

5.4 Overtopping volumes

An average overtopping discharge does not say much about the load of the dike or revetment caused by individual waves. The significance of the individual overtopping volumes can be shown from the example in Fig. 5.32, which gives the probability distribution function of individual overtopping volumes for an average overtopping discharge of 1.7 l/s per m, a wave period of $T_{m-1,0} = 5$ s and for 7 % of overtopping waves. In this Fig. 50 % of the overtopping waves result in an overtopping volume of less than 0.06 m³ per m width but 1 % of the overtopping waves result in an overtopping volume of more than 0.77 m³ per m width, which is more than 10 times larger.

The overtopping volumes per wave can be described by a Weibull distribution with a shape factor of 0.75 and a scale factor a. It is a sharply upward bound curve in Fig. 5.32, showing that only a few very large overtopping waves count for most of the overtopping discharge. The shape factor was found to be almost constant. The scale factor a depends on the average overtopping rate q, the mean wave period T_m and the probability of overtopping waves P_{ov}. The Weibull distribution giving the exceedance probability P_V of an overtopping volume per wave V is described as:

$$P_V = P(\underline{V} \leq V) = 1 - \exp\left[-\left(\frac{V}{a}\right)^{0.75}\right] \qquad 5.32$$

with:

$$a = 0.84 \cdot T_m \cdot \frac{q}{P_{ov}} \qquad 5.33$$

If the overtopping volume per wave for a given probability of exceedance P_V is required:

$$V = a \left[- \ln (1 - P_V)\right]^{4/3} \qquad 5.34$$

For the maximum overtopping volume in a storm the following formula can be used, by filling in the number of overtopping waves N_{ov}. Note that the prediction of this maximum volume is subject to quite some uncertainty, which is always the case for a maximum in a distribution.

$$V_{max} = a \cdot \left[\ln (N_{ov})\right]^{4/3} \qquad 5.35$$

The probability of overtopping per wave can be calculated by assuming a Rayleigh-distribution of the wave run-up heights and taking $R_{u2\%}$ as a basis:

$$P_{ov} = \exp\left[-\left(\sqrt{-\ln 0.02}\,\frac{R_C}{R_{u2\%}}\right)^2\right] \qquad 5.36$$

The probability of overtopping per wave P_{ov} is related to the number of incoming (N_w) and overtopping waves (N_{ow}) by:

$$P_{ov} = \frac{N_{ow}}{N_w} \qquad 5.37$$

Example:
The probability distribution function for wave overtopping volumes per wave is calculated for a smooth $\tan\alpha$ = 1:6 dike with a freeboard of R_C = 2.0 m, a period of the incoming wave of $T_{m-1,0}$ = 5.0 s and a wave height of the incoming waves of H_{m0} = 2.0 m. For these conditions, the wave run-up height is $R_{u2\%}$ = 2.43 m, the average overtopping rate q = 1.7 l/(sm) and the probability of overtopping per wave is P_{ov} = 0.071. This means, that the scale factor a becomes a = 0.100. The storm duration is assumed to be 1 hour, resulting in 720 incoming waves and 51 overtopping waves. The probability of exceedance curve is given in Fig. 5.32.

5.5 Overtopping flow velocities and overtopping flow depth

Average overtopping rates are not appropriate to describe the interaction between the overtopping flow and the failure mechanisms (infiltration and erosion) of a clay dike. Therefore, research was carried out recently in small and large scale model tests to investigate the overtopping flow (see Fig. 5.33) velocities and the related flow depth on the seaward slope, the dike crest and the landward slope. Results are summarized in SCHÜTTRUMPF and VAN GENT, 2003. Empirical and theoretical functions were derived and verified by experimental data in small and large scale. These parameters are required as boundary conditions for geotechnical investigations, such as required for the analysis of erosion, infiltration and sliding.

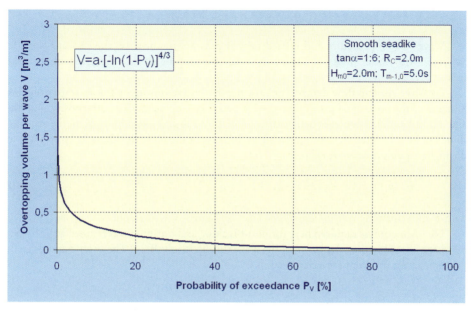

Fig. 5.32: Example probability distribution for wave overtopping volumes per wave

The parameters for overtopping flow velocities and overtopping flow depth will be described separately for the seaward slope, the dike crest and the landward slope.

Fig. 5.33: Wave overtopping on the landward side of a seadike (photo: ZITSCHER)

5.5.1 Seaward Slope

Wave run-up velocities and related flow depths are required on the seaward slope to determine the initial flow conditions of wave overtopping at the beginning of the dike crest.

(a) Wave run-up flow depth: The flow depth of wave run-up on the seaward slope is a function of the horizontal projection x_z of the wave run-up height $R_{u2\%}$, the position on the dike x_A and a dimensionless coefficient c_2. The flow depth of wave run-up on the seaward slope can be calculated by assuming a linear decrease of the layer thickness h_A from SWL to the highest point of wave run-up:

$$h_A(x_*) = c_2(x_z - x_A) = c_2 \cdot x_* \qquad 5.38$$

with x_* the remaining run-up length ($x_* = x_z - x_A$) and $x_z = R_{u2\%}/\tan\alpha$.

No distinction is required here for non-breaking and breaking waves since wave breaking is considered in the calculation of the wave run-up height $R_{u2\%}$. The coefficient c_2 can be determined for different exceedance levels by Table 5.3.

Table 5.3: Characteristic values for parameter c_2 (TMA-spectra)

Parameter	c_2	σ'
$h_{A,50\%}$	0,028	0.15
$h_{A,10\%}$	0,042	0.18
$h_{A,2\%}$	0,055	0.22

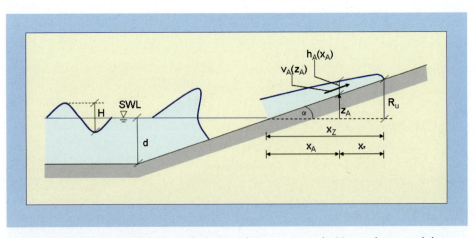

Fig. 5.34: Definition sketch for layer thickness and wave run-up velocities on the seaward slope

(b) Wave run-up velocities: The wave run-up velocity is defined as the maximum velocity that occurs during wave run-up at any position on the seaward slope. This velocity is attributed to the front velocity of the wave run-up tongue. The wave run-up velocity can be derived from a simplified energy equation and is given by:

$$v_A = k^* \cdot \sqrt{2g(R_{u2\%} - z_A)} \qquad 5.39$$

with v_A the wave run-up velocity at a point z_A above SWL, $R_{u2\%}$ the wave run-up height exceeded by 2 % of the incoming waves, and k^* a dimensionless coefficient.

In dimensionless form, the wave run-up velocity is:

$$\frac{v_A}{\sqrt{gH_S}} = a_0^* \sqrt{\frac{(R_{u2\%} - z_A)}{H_S}} \qquad 5.40$$

Equation 5.40 has been calibrated by small and large scale model data resulting in values for the 2 %, 10 % and 50 % exceedance probability (Table 5.4).

Exemplarily, the decrease of wave run-up velocity and wave run-up flow depth on the seaward slope is given in Fig. 5.35.

Table 5.4: Characteristic Values for Parameter a_0^* (TMA-spectra)

Parameter	a_0^*	σ'
$v_{A,50\%}$	1.03	0.23
$v_{A,10\%}$	1.37	0.18
$v_{A,2\%}$	1.55	0.15

5.5.2 Dike Crest

The overtopping tongue arrives as a very turbulent flow at the dike crest (Fig. 5.36). The water is full of air bubbles and the flow can be called "white water flow". Maximum flow depth and overtopping velocities were measured in this overtopping phase over the crest. The overtopping flow separates slightly from the dike surface at the front edge of the crest. No flow separation occurs at the middle and at the rear edge of the crest. In the second overtopping phase, the overtopping flow has crossed the crest. Less air is in the overtopping flow but the flow itself is still very turbulent with waves in flow direction and normal to flow direction. In the third overtopping phase, a second peak arrives at the crest resulting in nearly the same flow depth as the first peak. In the fourth overtopping phase, the air has disappeared from the overtopping flow and both overtopping velocity and flow depth are decreasing. Finally, the overtopping flow nearly stops on the dike crest for small overtopping flow depths. Few air is in the overtopping water. At the end of this phase, the overtopping water on the dike crest starts flowing seaward.

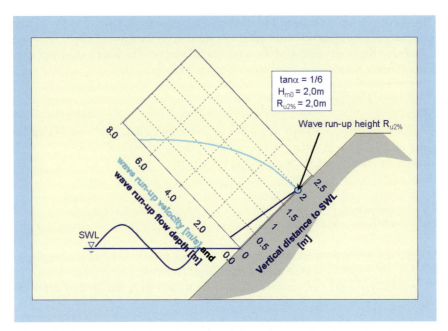

Fig. 5.35: Wave run-up velocity and wave run-up flow depth on the seaward slope (example)

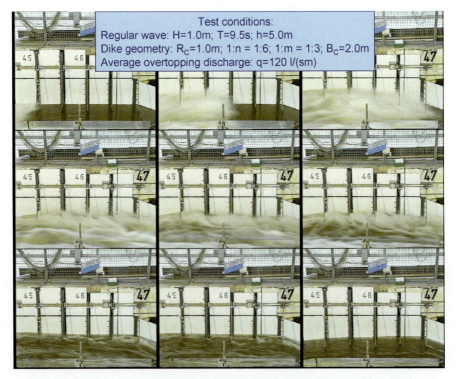

Fig. 5.36: Sequence showing the transition of overtopping flow on a dike crest (Large Wave Flume, Hannover)

The flow parameters at the transition line between seaward slope and dike crest are the initial conditions for the overtopping flow on the dike crest. The evolution of the overtopping flow parameters on the dike crest will be described below.

(a) **Overtopping flow depth on the dike crest:** The overtopping flow depth on the dike crest depends on the width of the crest B and the co-ordinate on the crest x_C (Fig. 5.37). The overtopping flow depth on the dike crest decreases due to the fact that the overtopping water is deformed. Thus, the decrease of overtopping flow depth over the dike crest can be described by an exponential function:

$$\frac{h_C(x_C)}{h_C(x_C=0)} = \frac{c_2(x_C)}{c_2(x_C=0)} = \exp\left(-c_3 \frac{x_C}{B_C}\right) \qquad 5.41$$

with h_C the overtopping flow depth on the dike crest, x_C the horizontal coordinate on the dike crest with $x_C = 0$ at the beginning of the dike crest, c_3 the dimensional coefficient = 0.89 for TMA spectra (σ' = 0.06) and 1.11 for natural wave spectra (σ' = 0.09), and B_C the width of the dike crest (for B_c = 2 to 3 m in prototype scale).

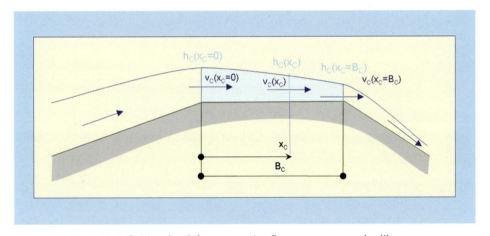

Fig. 5.37: Definition sketch for overtopping flow parameters on the dike crest

(b) **Overtopping flow velocity:** A theoretical function for overtopping flow velocities on the dike crest has been developed by using the simplified Navier-Stokes-equations and the following assumptions: the dike crest is horizontal; velocities vertical to the dike slope can be neglected; the pressure term is almost constant over the dike crest; viscous effects in flow direction are small; bottom friction is constant over the dike crest.

The following formula was derived from the Navier-Stokes-equations and verified by small and large scale model tests (Fig. 5.38):

$$v_C = v_{C(x_C=0)} \exp\left(-\frac{x_C \cdot f}{2 \cdot h_C}\right) \qquad 5.42$$

with v_C the overtopping flow velocity on the dike crest; $v_{C, (x_C = 0)}$ the overtopping flow velocity at the beginning of the dike crest ($x_C = 0$); x_C the coordinate along the dike crest; f the friction coefficient; and h_C the flow depth at x_C.

From Equation 5.43 it is obvious that the overtopping flow velocity on the dike crest is mainly influenced by bottom friction. The overtopping flow velocity decreases from the beginning of the dike crest to the end of the dike crest due to bottom friction. The friction factor f was determined from model tests at straight and smooth slope to be f = 0.01. The importance of the friction factor on the overtopping flow velocities on the dike crest is obvious from Fig. 5.39. The overtopping flow velocity decreases significantly over the dike crest for increasing surface roughness. But for flow depths larger than about 0.1 m and dike crest widths around 2–3 m, the flow depth and velocity hardly change over the crest.

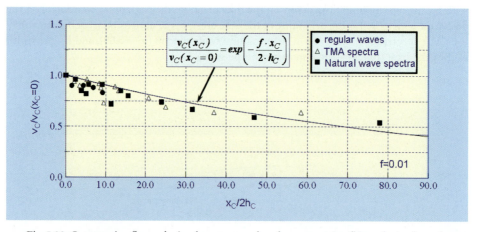

Fig. 5.38: Overtopping flow velocity data compared to the overtopping flow velocity formula

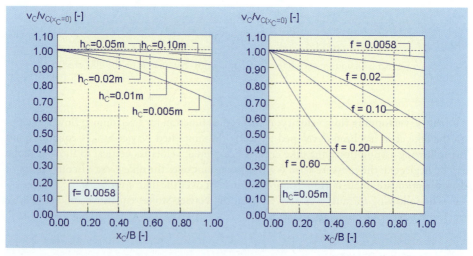

Fig. 5.39: Sensitivity analysis for the dike crest (left side: influence of overtopping flow depth on overtopping flow velocity; right side: influence of bottom friction on overtopping flow velocity)

5.5.3 Landward Slope

The overtopping water flows from the dike crest to the landward slope of the dike. The description of the overtopping process on the landward slope is very important with respect to dike failures which often occurred on the landward slope in the past. An analytical function was developed which describes overtopping flow velocities and overtopping flow depths on the landward slope as a function of the overtopping flow velocity at the end of the dike crest ($v_{b,0} = v_C(x_C = B)$), the slope angle β of the landward side and the position s_B on the landward side with $s_B = 0$ at the intersection between dike crest and landward slope. A definition sketch is given in Fig. 5.41. The following assumptions were made to derive an analytical function from the Navier-Stokes-equations: velocities vertical to the dike slope can be neglected; the pressure term is almost constant over the dike crest; and the viscous effects in flow direction are small.

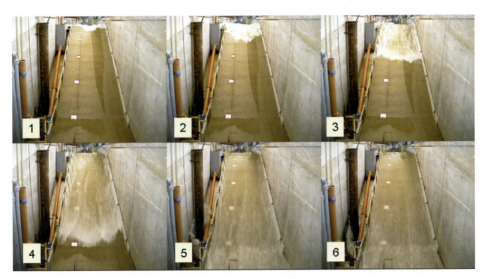

Fig. 5.40: Overtopping flow on the landward slope (Large Wave Flume, Hannover) (photo: SCHÜTTRUMPF)

This results in the following formula for overtopping flow velocities:

$$v_b = \frac{v_{b,0} + \dfrac{k_1 h_b}{f} \tanh\left(\dfrac{k_1 t}{2}\right)}{1 + \dfrac{f\, v_{b,0}}{h_b k_1} \tanh\left(\dfrac{k_1 t}{2}\right)}$$

5.43

with:

$$t \approx -\frac{v_{b,0}}{g \sin\beta} + \sqrt{\frac{v_b^2}{g^2 \sin^2\beta} + \frac{2 s_b}{g \sin\beta}} \quad \text{and} \quad k_1 = \sqrt{\frac{2 f g \sin\beta}{h_b}}$$

Equation 5.44 needs an iterative solution since the overtopping flow depth h_b and the overtopping flow velocity v_b on the landward slope are unknown. The overtopping flow depth h_b can be replaced in a first step by:

$$h_b = \frac{v_{b,0} \cdot h_{b,0}}{v_b} \qquad 5.44$$

with $v_{b,0}$ the overtopping flow velocity at the beginning of the landward slope ($v_{b,0} = v_B(s_B = 0)$); and $h_{b,0}$ the overtopping flow depth at the beginning of the landward slope ($h_{b,0} = h_B(s_B = 0)$).

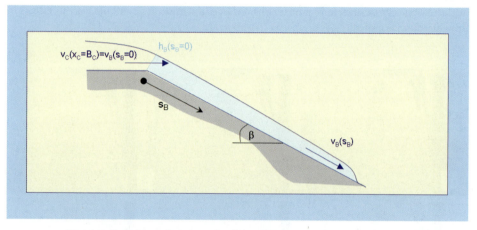

Fig. 5.41: Definition of overtopping flow parameters on the landward slope

In Fig. 5.42, the influence of the landward slope on overtopping flow velocities and overtopping flow depths is shown. The landward slope was varied between 1:m = 1:2 and 1:m = 1:6 which is in the practical range. It is obvious that overtopping flow velocities increase for steeper slopes and related overtopping flow depths decrease with increasing slope steepness.

The second important factor influencing the overtopping flow on the landward slope is the bottom friction coefficient f which has to be determined experimentally. Some references for the friction coefficient on wave run-up are given in literature (e.g. VAN GENT, 1995; CORNETT and MANSARD, 1994, SCHULZ, 1992). Here, the bottom friction coefficient was determined by comparison of the experimental to be f = 0.02 for a smooth and straight slope. These values are comparable to references in literature. VAN GENT (1995) recommends a friction coefficient f = 0.02 for smooth slopes and SCHULZ (1992) determined friction coefficients between 0.017 and 0.022.

The overtopping flow on the landward slope tends towards an asymptote for $s_b \to \infty$ which is given by:

$$v_b = \sqrt{\frac{2 \cdot g \cdot h_b \cdot \sin \beta}{f}} \qquad 5.45$$

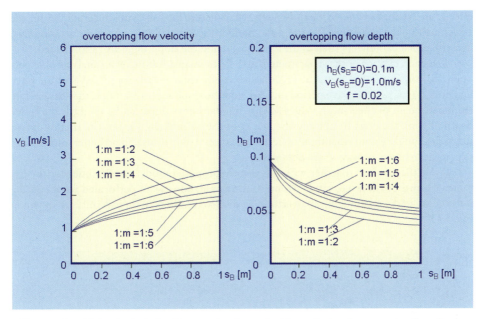

Fig. 5.42: Sensitivity Analysis for Overtopping flow velocities and related overtopping flow depths
– Influence of the landward slope –

5.6 Scale effects for dikes

A couple of investigations on the influence of wind and scale effects are available for sloping structures all of which are valid only for rough structures. Sea dikes are generally smooth and covered e.g. by grass, revetment stones or asphalt which all have roughness coefficients larger than $\gamma_f = 0.9$. Hence, there are no significant scale effects for these roughness coefficients. This is however only true if the model requirements as given in Table 4.3 in Section 4.8.3 are respected.

For rough slopes as they e.g. occur for any roughness elements on the seaward slope, scale effects for low overtopping rates cannot be excluded and therefore, the procedure as given in Section 6.3.6 should be applied.

5.7 Uncertainties

In section 5.3.1 model uncertainties have been introduced in the calculation by defining the parameter b in Equation 5.8 as normally distributed parameter with a mean of 4.75 and a standard deviation of $\sigma = 0.5$ for breaking waves and b = 0.2 and $\sigma = 0.35$ for non breaking waves. This has also been illustrated by Fig. 5.6 and Fig. 5.7, respectively, showing the 90 % confidence interval resulting from these considerations.

In using the approach as proposed in section 4.8.1, a model uncertainty of about 60 % is obtained. Note that this approach comprises a model factor for Equation 5.8 in total rather than the uncertainty of the parameter b only as used in Fig. 5.6 and Fig. 5.7. The latter approach comprise various uncertainties from model tests, incl. repeatability of tests, model effects, uncertainties in wave measurements, etc. whereas the following uncertainties for the assessment of the wave heights, the wave period, the water depth, the wave attack angle, constructional parameters such as the crest height and the slope angle are not included.

The uncertainties of these parameters may be estimated following an analysis of expert opinions from SCHÜTTRUMPF et al. (2006) using coefficients of variations (CoV) for the wave height H_{m0} (3.6 %), the wave period (4.0 %), and the slope angle (2.0 %). Other parameters are independent of their mean values so that standard deviations can be used for the water depth (0.1 m), the crest height and the height of the berm (0.06 m), and the friction factor (0.05). It should be noted that these uncertainties should only be used if no better information (e.g. measurements of waves) are obtainable.

Using these values together with the already proposed model uncertainties for the parameter b in Equation 5.8, crude Monte Carlo simulations were performed to obtain the uncertainty in the resulting mean overtopping discharges. Plots of these results are shown in Fig. 5.43.

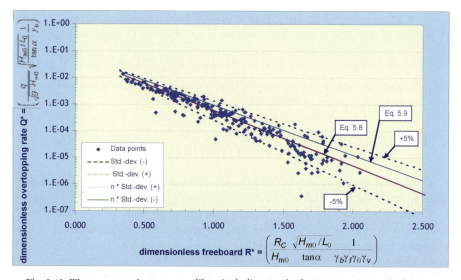

Fig. 5.43: Wave overtopping over sea dikes, including results from uncertainty calculations

As compared to Fig. 5.6 and Fig. 5.7, respectively, it can be seen that the resulting curves (denominated as 'n*std.-dev.' in Fig. 5.43) are only giving slightly larger uncertainty bands as the 5% lines resulting from calculations with model uncertainties. This suggests a very large influence of the model uncertainties so that no other uncertainties, if assumed to be in the range as given above, need to be considered. It is therefore proposed to use Equations 5.8 and 5.9 as suggested in section 5.3.1. In case of deterministic calculations, Equation 5.9 should be used with no further adaptation of parameters. In case of probabilistic calculations, Equation 5.8 should be used and uncertainties of all input parameters should be considered in addition to the model uncertainties. If detailed information of some of these parameters is not available, the uncertainties as proposed above may be used.

It should be noted that only uncertainties for mean wave overtopping rates are considered here. Other methods such as flow velocities and flow depths were not considered here but can be dealt with using the principal procedure as discussed in section 1.5.4.

6. Armoured rubble slopes and mounds

6.1 Introduction

This manual describes three types of flood defences or coastal structures:
- coastal dikes and embankment seawalls,
- armoured rubble slopes and structures,
- and vertical and steep seawalls.

Sometimes there will be combinations and it will be difficult to place them only in one category. For example, a vertical wall or sloping embankment with a large rock berm in front. Armoured rubble slopes and mounds () are characterized by a mound with some porosity or permeability, covered by a sloping porous armour layer consisting of large rock or concrete units. In contrast to dikes and embankment seawalls the porosity of the structure and armour layer plays a role in wave run-up and overtopping. The cross-section of a rubble mound slope, however, may have great similarities with an embankment seawall and may consist of various slopes.

As rubble mound structures are to some extent similar to dikes and embankment seawalls, the basic wave run-up and overtopping formulae are taken from Chapter 5. They will then be modified, if necessary, to fit for rubble mound structures. Also for most definitions the reader is referred to Chapter 5 (or Chapter 1.4). More in particular:
- the definition of wave run-up (Fig. 5.3)
- the general wave run-up formula (Equation 5.1)
- the general wave overtopping formula (Equation 5.8 or 5.9)
- the influence factors γ_b, γ_f and γ_β
- the spectral wave period $T_{m-1,0}$
- the difference in deterministic and probabilistic approach

The main calculation procedure for armoured rubble slopes and mounds is given in Table 6.1.

Table 6.1: Main calculation procedure for armoured rubble slopes and mounds

	Deterministic design	Probabilistic design
Wave run-up height (2%)	Eq. 6.2	Eq. 6.1
Wave runu-up height for shingle beaches		Eq. 6.20
Mean wave overtopping discharge	Eq. 6.5	Eq. 6.6
Mean overtopping discharge for berm breakwaters		Eq. 6.9 – 6.11
Percentage of overtopping waves		Eq. 6.4
Individual overtopping volumes	Eqs. 6.15-6.16	Eqs. 6.15-6.16
Effect of armour roughness	Table 6.2	Table 6.2
Effect of armour crest berm	Eq. 6.7	Eq. 6.7
Effect of oblique waves	Eq. 6.8 for overtopping	Eq. 6.8 for overtopping
Overtopping velocities		Eqs. 6.17 – 6.18
Scale and model uncertenties	Eqs. 6.12 – 6.14	Eqs. 6.12 – 6.14

Fig. 6.1: Armoured structures

6.2 Wave run-up and run-down levels, number of overtopping waves

Through civil engineering history the wave run-up and particularly the 2% run-up height was important for the design of dikes and coastal embankments. Till quite recently the 2% run-up height under design conditions was considered a good measure for the required dike height. With only 2% of overtopping waves the load on crest and inner side were considered so small that no special measurements had to be taken with respect to strength of these parts of a dike. Recently, the requirements for dikes changed to allowable wave overtopping, making the 2% run-up value less important in engineering practice.

Wave run-up has always been less important for rock slopes and rubble mound structures and the crest height of these type of structures has mostly been based on allowable overtopping, or even on allowable transmission (low-crested structures). Still an estimation or prediction of wave run-up is valuable as it gives a prediction of the number or percentage of waves which will reach the crest of the structure and eventually give wave overtopping. And this number is needed for a good prediction of individual overtopping volumes per wave.

Fig. 6.2 gives 2% wave run-up heights for various rocks slopes with $\cot\alpha$ = 1.5, 2, 3 and 4 and for an impermeable and permeable core of the rubble mound. These run-up measurements were performed during the stability tests on rock slopes of VAN DER MEER (1988). First of all the graph gives values for a large range of the breaker parameter $\xi_{m-1,0}$, due to the fact that various slope angels were tested, but also with long wave periods (giving large $\xi_{m-1,0}$-values). Most breakwaters have steep slopes 1:1.5 or 1:2 only and then the range of breaker parameters is often limited to $\xi_{m-1,0}$ = 2-4. The graph gives rock slope information outside this range, which may be useful also for slopes with concrete armour units.

The highest curve in Fig. 6.2 gives the prediction for smooth straight slopes, see Fig. 5.1 and Equation 5.3. A rubble mound slope dissipates significantly more wave energy than an equivalent smooth and impermeable slope. Both the roughness and porosity of the armour layer cause this effect, but also the permeability of the under layer and core contribute to it. Fig. 6.2 shows the data for an impermeable core (geotextile on sand or clay underneath a thin under layer) and for a permeable core (such as most breakwaters). The difference is most significant for large breaker parameters.

Equation 5.1 includes the influence factor for roughness γ_f. For two layers of rock on an impermeable core γ_f = 0.55. This reduces to γ_f = 0.40 for two layers of rock on a permeable core. This influence factor is used in the linear part of the run-up formula, say for $\xi_0 \leq 1.8$. From $\xi_{m-1,0}$ = 1.8 the roughness factor increases linearly up to 1 for $\xi_{m-1,0}$ = 10 and it remains 1 for larger values. For a permeable core, however, a maximum is reached for $R_{u2\%}/H_{m0}$ = 1.97. The physical explanation for this is that if the slope becomes very steep (large ξ_0-value) and the core is impermeable, the surging waves slowly run up and down the slope and all the water stays in the armour layer, leading to fairly high run-up. The surging wave actually does not "feel" the roughness anymore and acts as a wave on a very steep smooth slope. For an permeable core, however, the water can penetrate into the core which decreases the actual run-up to a constant maximum (the horizontal line in Fig. 6.2).

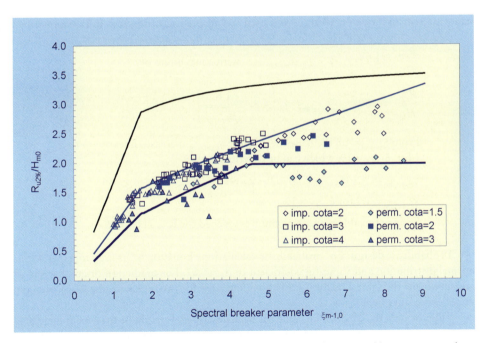

Fig. 6.2: Relative run-up on straight rock slopes with permeable and impermeable core, compared to smooth impermeable slopes

The prediction for the 2% mean wave run-up value for rock or rough slopes can be described by:

$$\frac{R_{u2\%}}{H_{m0}} = 1.65 \cdot \gamma_b \cdot \gamma_f \cdot \gamma_\beta \cdot \xi_{m-1,0} \quad \text{with a maximum of}$$

$$\frac{R_{u2\%}}{H_{m0}} = 1.00 \cdot \gamma_b \cdot \gamma_{f\,surging} \cdot \gamma_\beta \left(4.0 - \frac{1.5}{\sqrt{\xi_{m-1,0}}}\right)$$

6.1

From $\xi_{m-1,0} = 1.8$ the roughness factor $\gamma_{f\,surging}$ increases linearly up to 1 for $\xi_{m-1,0} = 10$, which can be described by:

$\gamma_{f\,surging} = \gamma_f + (\xi_{m-1,0} - 1.8) * (1 - \gamma_f)/8.2$

$\gamma_{f\,surging} = 1.0$ for $\xi_{m-1,0} > 10$.

For a permeable core a maximum is reached for $R_{u2\%}/H_{m0} = 1.97$

Equation 6.1 may also give a good prediction for run-up on slopes armoured with concrete armour units, if the right roughness factor is applied (see Section 6.3).

Deterministic design or safety assessment: For design or a safety assessment of the crest height, it is advised not to follow the average trend, but to include the uncertainty of the prediction, see Section 5.2. As the basic equation is similar for a smooth and a rough slope, the method to include uncertainty is also the same. This means that for a deterministic design or safety assessment Equation 5.4 should be used and adapted accordingly as in Equation 6.1:

$$\frac{R_{u2\%}}{H_{m0}} = 1.75 \cdot \gamma_b \cdot \gamma_f \cdot \gamma_\beta \cdot \xi_{m-1,0} \quad \text{with a maximum of}$$

$$\frac{R_{u2\%}}{H_{m0}} = 1.00 \cdot \gamma_b \cdot \gamma_{f\,surging} \cdot \gamma_\beta \left(4.3 - \frac{1.6}{\sqrt{\xi_{m-1,0}}}\right)$$

6.2

From $\xi_{m-1,0} = 1.8$ the roughness factor $\gamma_{f\,surging}$ increases linearly up to 1 for $\xi_{m-1,0} = 10$, which can be described by:

$\gamma_{f\,surging} = \gamma_f + (\xi_{m-1,0} - 1.8) * (1 - \gamma_f)/8.2$

$\gamma_{f\,surging} = 1.0$ for $\xi_{m-1,0} > 10$.

For a permeable core a maximum is reached for $R_{u2\%}/H_{m0} = 2.11$

Probabilistic design: For probabilistic calculations Equation 6.1 is used together with a normal distribution and variation coefficient of $\sigma' = 0.07$. For prediction or comparison of measurements the same Equation 6.1 is used, but now for instance with the 5% lower and upper exceedance lines.

Till now only the 2% run-up value has been described. It might be that one is interested in an other percentage, for example for design of breakwaters where the crest height may be determined by an allowable percentage of overtopping waves, say 10–15%. A few ways exist to calculate run-up heights for other percentages, or to calculate the number of overtopping waves for a given crest height. VAN DER MEER and STAM (1992) give two methods. One is an equation like 6.1 with a table of coefficients for the 0.1%, 1%, 2%, 5%, 10% and 50% (median). Interpolation is needed for other percentages.

The second method gives a formula for the run-up distribution as a function of wave conditions, slope angle and permeability of the structure. The distribution is a two-parameter Weibull distribution. With this method the run-up can be calculated for every percentage wanted. Both methods apply to straight rock slopes only and will not be described here. The given references, however, give all details.

The easiest way to calculate run-up (or overtopping percentage) different from 2 % is to take the 2%-value and assume a Rayleigh distribution. This is similar to the method in Chapter 5 for dikes and embankment seawalls. The probability of overtopping $P_{ov} = N_{ow}/N_w$ (the percentage is simply 100 times larger) can be calculated by:

$$P_{ov} = N_{ow} / N_w = \exp\left[-\left(\sqrt{-\ln 0.02}\, \frac{R_C}{R_{u,2\%}}\right)^2\right] \qquad 6.3$$

Equation 6.3 can be used to calculate the probability of overtopping, given a crest freeboard R_c or to calculate the required crest freeboard, given an allowable probability or percentage of overtopping waves.

One warning should be given in applying Equations 6.1, 6.2 and 6.3. The equations give the run-up level in percentage or height on a straight (rock) slope. This is not the same as the number of overtopping waves or overtopping percentage. Fig. 6.3 gives the difference. The run-up is always a point on a straight slope, where for a rock slope or armoured mound the overtopping is measured some distance away from the seaward slope and on the crest, often behind a crown wall. This means that Equations 6.1, 6.2 and 6.3 always give an over estimation of the number of overtopping waves.

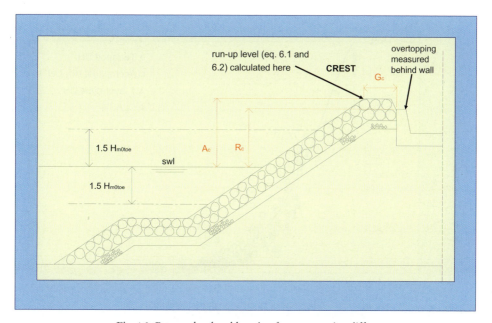

Fig. 6.3: Run-up level and location for overtopping differ

Fig. 6.4 shows measured data for rubble mound breakwaters armoured with Tetrapods (DE JONG 1996), Accropode™ or a single layer of cubes (VAN GENT et al. 1999). All tests were performed at Delft Hydraulics. The test set-up was more or less similar to Fig. 6.2 with a crown wall height R_c a little lower than the armour freeboard A_c. CLASH-data on specific overtopping tests (see Section 6.3) for various rock and concrete armoured slopes were added to Fig. 6.4. This Fig. gives only the percentage of overtopping waves passing the crown wall. Analysis showed that the size of the armour unit relative to the wave height had influence, which gave a combined parameter $A_c{*}D_n/H_{m0}^2$, where D_n is the nominal diameter of the armour unit.

The Fig. covers the whole range of overtopping percentages, from complete overtopping with the crest at or lower than SWL to no overtopping at all. The CLASH data give maximum overtopping percentages of about 30 %. Larger percentages mean that overtopping is so large that it can hardly be measured and that wave transmission starts to play a role.

Taking 100 % overtopping for zero freeboard (the actual data are only a little lower), a Weibull curve can be fitted through the data. Equation 6.4 can be used to predict the number or percentage of overtopping waves or to establish the armour crest level for an allowable percentage of overtopping waves.

$$P_{ov} = N_{ow}/N_w = \exp\left[-\left(\frac{A_c D_n}{0.19 H_{m0}^2}\right)^{1.4}\right] \qquad 6.4$$

It is clear that equations 6.1–6.3 will come to more overtopping waves than equation 6.4. But both estimations together give a designer enough information to establish the required crest height of a structure given an allowable overtopping percentage.

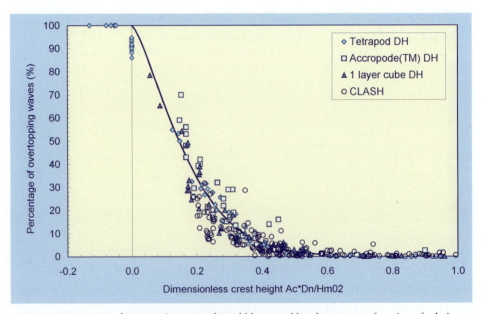

Fig. 6.4: Percentage of overtopping waves for rubble mound breakwaters as a function of relative (armour) crest height and armour size ($R_c \leq A_c$)

When a wave on a structure has reached its highest point it will run down on the slope till the next wave meets this water and run-up starts again. The lowest point to where the water retreats, measured vertically to SWL, is called the run-down level. Run-down often is less or not important compared to wave run-up, but both together they may give an idea of the total water excursion on the slope. Therefore, only a first estimate of run-down on straight rock slopes is given here, based on the same tests of VAN DER MEER (1988), but re-analysed with respect to the use of the spectral wave period $T_{m-1,0}$. Fig. 6.5 gives an overall view.

The graph shows clearly the influence of the permeability of the structure as the solid data points (impermeable core) generally show larger run-down than the open data symbols of the permeable core. Furthermore, the breaker parameter $\xi_{m-1,0}$ gives a fairly clear trend of run-down for various slope angles and wave periods. Fig. 6.5 can be used directly for design purposes, as it also gives a good idea of the scatter.

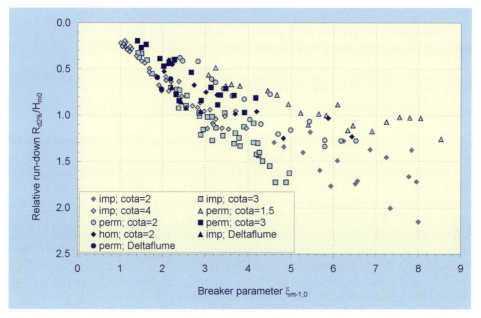

Fig. 6.5: Relative 2 % run-down on straight rock slopes with impermeable core (imp), permeable core (perm) and homogeneous structure (hom)

6.3 Overtopping discharges

6.3.1 Simple armoured slopes

The mean overtopping discharge is often used to judge allowable overtopping. It is easy to measure and an extensive database on mean overtopping discharge has been gathered in CLASH. This mean discharge does of course not describe the real behaviour of wave overtopping, where only large waves will reach the top of the structure and give overtopping. Random individual wave overtopping means random in time and each wave gives a different overtopping volume. But the description of individual overtopping is based on the mean overtopping, as the duration of overtopping multiplied with this mean overtopping discharge gives the total volume of water overtopped by a certain number of overtopping waves. The mean overtopping discharge has been described in this section. The individual overtopping volumes is the subject in Section 6.4

Just like for run-up, the basic formula for mean wave overtopping discharge has been described in Chapter 5 for smooth slopes (Equation 5.8 or 5.9). The influence factor for roughness should take into account rough structures. Rubble mound structures often have steep slopes of about 1:1.5, leading to the second part in the overtopping equations.

Deterministic design or safety assessment: The equation, including a standard deviation of safety, should be used for deterministic design or safety assessment:

$$\frac{q}{\sqrt{g \cdot H_{m0}^3}} = 0.2 \cdot \exp\left(-2.3 \frac{R_C}{H_{m0} \cdot \gamma_f \cdot \gamma_\beta}\right) \qquad 6.5$$

Probabilistic design: The mean prediction should be used for probabilistic design, or prediction of or comparison with measurements. This equation is given by:

$$\frac{q}{\sqrt{g \cdot H_{m0}^3}} = 0.2 \cdot \exp\left(-2.6 \frac{R_C}{H_{m0} \cdot \gamma_f \cdot \gamma_\beta}\right) \qquad 6.6$$

The coefficient 2.6 in Equation 6.6 gives the mean prediction and its reliability can be described by a standard deviation of $\sigma = 0.35$.

As part of the EU research programme CLASH (BRUCE et al. 2007) tests were undertaken to derive roughness factors for rock slopes and different armour units on sloping permeable structures. Overtopping was measured for a 1:1.5 sloping permeable structure at a reference point $3D_n$ from the crest edge, where D_n is the nominal diameter. The wave wall had the same height as the armour crest, so $R_c = A_c$. As discussed in Section 6.2 and Fig. 6.3, the point to where run-up can be measured and the location of overtopping may differ. Normally, a rubble mound structure has a crest width of at least $3D_n$. Waves rushing up the slope reach the crest with an upward velocity. For this reason it is assumed that overtopping waves reaching the crest, will also reach the location $3D_n$ further.

Results of the CLASH-work is shown in Fig. 6.6 and Table 6.2. Fig. 6.6 gives all data together in one graph. Two lines are given, one for a smooth slope, Equation 6.4 with $\gamma_f = 1.0$, and one for rubble mound 1:1.5 slopes, with the same equation, but with $\gamma_f = 0.45$. The lower line only gives a kind of average, but shows clearly the very large influence of roughness and permeability on wave overtopping. The required crest height for a steep rubble mound structure is at least half of that for a steep smooth structure, for similar overtopping discharge. It is also for this reason that smooth slopes are often more gentle in order to reduce the crest heights.

In Fig. 6.6 one-layer systems, like Accropode™, CORE-LOC®, Xbloc® and 1 layer of cubes, have solid symbols. Two-layer systems have been given by open symbols. There is a slight tendency that one-layer systems give a little more overtopping than two-layer systems, which is also clear from Table 6.2. Equation 6.4 can be used with the roughness factors in Table 6.2 for prediction of mean overtopping discharges for rubble mound breakwaters. Values in italics in Table 6.2 have been estimated/extrapolated, based on the CLASH results.

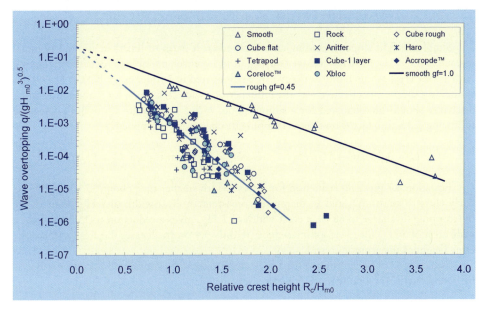

Fig. 6.6: Mean overtopping discharge for 1:1.5 smooth and rubble mound slopes

Table 6.2: Values for roughness factor γ_f for permeable rubble mound structures with slope of 1:1.5. Values in italics are estimated/extrapolated

Type of armour layer	γ_f
Smooth impermeable surface	1.00
Rocks (1 layer, impermeable core)	0.60
Rocks (1 layer, permeable core)	0.45
Rocks (2 layers, impermeable core)	0.55
Rocks (2 layers, permeable core)	0.40
Cubes (1 layer, random positioning)	0.50
Cubes (2 layers, random positioning)	0.47
Antifers	0.47
HARO's	0.47
Accropode™	0.46
Xbloc®	0.45
CORE-LOC®	0.44
Tetrapods	0.38
Dolosse	*0.43*

6.3.2 Effect of armoured crest berm

Simple straight slopes including an armoured crest berm of less than about 3 nominal diameters ($G_c \approx 3D_n$) will reduce overtopping. It is, however, possible to reduce overtopping with a wide crest as much more energy can be dissipated in a wider crest. BESLEY (1999) describes in a simple and effective way the influence of a wide crest. First the wave overtopping discharge should be calculated for a simple slope, with a crest width up to $3D_n$. Then the following reduction factor on the overtopping discharge can be applied:

$$C_r = 3.06\exp(-1.5G_c/H_{m0}) \qquad G_c/H_{m0} \quad \text{with maximum} \quad C_r = 1 \qquad 6.7$$

Equation 6.7 gives no reduction for a crest width smaller than about $0.75\ H_{m0}$. This is fairly close to about $3D_n$ and is, therefore, consistent. A crest width of $1\ H_{m0}$ reduces the overtopping discharge to 68 %, a crest width of $2\ H_{m0}$ gives a reduction to 15 % and for a wide crest of $3\ H_{m0}$ the overtopping reduces to only 3.4 %. In all cases the crest wall has the same height as the armour crest: $R_c = A_c$.

Equation 6.7 was determined for a rock slope and can be considered as conservative, as for a slope with Accropode more reduction was found.

6.3.3 Effect of oblique waves

Section 5.5.3 describes the effect of oblique waves on run-up and overtopping on smooth slopes (including some roughness). But specific tests on rubble mound slopes were not performed at that time. In CLASH, however, this omission was discovered and specific tests on a rubble mound breakwater were performed with a slope of 1:2 and armoured with rock or cubes (ANDERSEN and BURCHARTH, 2004). The structure was tested both with long-crested and short-crested waves, but only the results by short-crested waves will be given.

For oblique waves the angle of wave attack β (deg.) is defined as the angle between the direction of propagation of waves and the axis perpendicular to the structure (for perpendicular wave attack: β = 0°). And the direction of wave attack is the angle after any change of direction of the waves on the foreshore due to refraction. Just like for smooth slopes, the influence of the angle of wave attack is described by the influence factor γ_β. Just as for smooth slopes there is a linear relationship between the influence factor and the angle of wave attack, but the reduction in overtopping is much faster with increasing angle:

$$\gamma_\beta = 1 - 0.0063|\beta| \qquad \text{for } 0° \leq |\beta| \leq 80°$$
$$\text{for } |\beta| > 80° \text{ the result } \beta = 80° \text{ can be applied} \qquad 6.8$$

The wave height and period are linearly reduced to zero for $80° \leq |\beta| \leq 110°$, just like for smooth slopes, see Section 5.3.3. For $|\beta| > 110°$ the wave overtopping is assumed to be q = 0 m³/s/m.

6.3.4 Composite slopes and berms, including berm breakwaters

In every formula where a cotα or breaker parameter $\xi_{m-1,0}$ is present, a procedure has to be described how a composite slope has to be taken into account. Hardly any specific research exists for rubble mound structures and, therefore, the procedure for composite slopes at sloping impermeable structures like dikes and sloping seawalls is assumed to be applicable. The procedure has been described in Section 5.3.4.

Also the influence of a berm in a sloping profile has been described in Section 5.3.4 and can be used for rubble mound structures. There is, however, often a difference in effect of composite slopes or berms for rubble mound and smooth gentle slopes. On gentle slopes the breaker parameter $\xi_{m-1,0}$ has large influence on wave overtopping, see Equations 5.8 and 5.9 as the breaker parameter will be quite small. Rubble mound structures often have a steep slope, leading to the formula for "non-breaking" waves, Equations 6.5 and 6.6. In these equations there is no form factor present.

This means that a composite slope and even a, not too long, berm leads to the same overtopping discharge as for a simple straight rubble mound slope. Only when the average slope becomes so gentle that the maximum in Equations 5.8 or 5.9 does not apply anymore, then a berm and a composite slope will have effect on the overtopping discharge. Generally, average slopes around 1:2 or steeper do not show influence of the slope angle, or only to a limited extend.

A specific type of rubble mound structure is the berm breakwater (see Fig. 6.7). The original idea behind the berm breakwater is that a large berm, consisting of fairly large rock, is constructed into the sea with a steep seaward face. The berm height is higher than the

Fig. 6.7: Icelandic Berm breakwater

minimum required for construction with land based equipment. Due to the steep seaward face the first storms will reshape the berm and finally a structure will be present with a fully reshaped S-profile. Such a profile has then a gentle 1:4 or 1:5 slope just below the water level and steep upper and lower slopes, see Fig. 6.8.

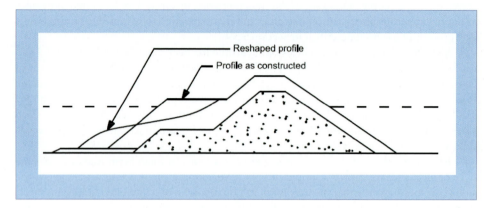

Fig. 6.8: Conventional reshaping berm breakwater

The idea of the reshaping berm breakwater has evolved in Iceland to a more or less non-reshaping berm breakwater. The main difference is that during rock production from the quarry care is taken to gather a few percent of really big rock. Only a few percent is required to strengthen the corner of the berm and part of the down slope and upper layer of the berm in such a way that reshaping will hardly occur. An example with various rock classes (class I being the largest) is given in Fig. 6.9. Therefore distinction has been made between conventional reshaping berm breakwaters and the non-reshaping Icelandic type berm breakwater.

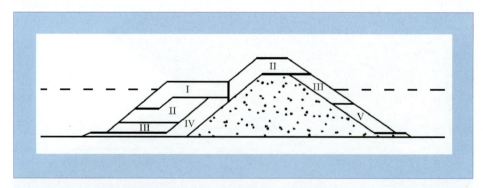

Fig. 6.9: Non-reshaping Icelandic berm breakwater with various classes of big rock

In order to calculate wave overtopping on reshaped berm breakwaters the reshaped profile should be known. The basic method of profile reshaping is given in VAN DER MEER (1988) and the programme BREAKWAT (WL | Delft Hydraulics) is able to calculate the profile. The first method described here to calculate wave overtopping at reshaping berm breakwaters is the method described in Chapter 5 (equations 5.8 or 5.9) with the roughness factors given in Table 6.1 of $\gamma_f = 0.40$ for reshaping berm breakwaters and $\gamma_f = 0.35$ for non-reshaping Icelandic berm breakwaters. The method of composite slopes and berms should be applied as described above.

The second method is to use the CLASH neural network (Section 4.4). As overtopping research at that time on berm breakwaters was limited, also this method gives quite some scatter, but a little less than the first method described above.

Recent information on berm breakwaters has been described by LYKKE ANDERSEN (2006). Only part of his research was included in the CLASH database and consequently in the Neural Network prediction method. He performed about 600 tests on reshaping berm breakwaters and some 60 on non-reshaping berm breakwaters (fixing the steep slopes by a steel net). The true non-reshaping Icelandic type of berm breakwaters with large rock classes, has not been tested and, therefore, his results might lead to an overestimation.

One comment should be made on the application of the results of LYKKE ANDERSEN (2006). The maximum overtopping discharge measured was only $q/(gH_{m0}^3)^{0.5} = 10^{-3}$. In practical situations with wave heights around 5 m the overtopping discharge will then be limited to only a few l/s per m width. For berm breakwaters and also for conventional rubble slopes and mounds allowable overtopping may be much higher than this value.

The final result of the work of LYKKE ANDERSEN (2006) is a quite complicated formula, based on multi-parameter fitting. The advantage of such a fitting is that by using a large number of parameters, the data set used will be quite well described by the formula. The disadvantage is that physical understanding of the working of the formula, certainly outside the ranges tested, is limited. But due to the fact that so many structures were tested, this effect may be negligible.

The formula is valid for berm breakwaters with no superstructure and gives the overtopping discharge at the back of the crest ($A_c = R_c$). In order to overcome the problem that one has to calculate the reshaped profile before any overtopping calculation can be done, the formula is based on the "as built" profile, before reshaping. Instead of calculating the profile, a part of the formula predicts the influence of waves on recession of the berm. The parameter used is called f_{H0}, which is an indicative measure of the reshaping and can be defined as a "factor accounting for the influence of stability numbers". Note that f_{H0} is a dimensionless factor and not the direct measure of recession and that H_0 and T_0 are also dimensionless parameters.

$$f_{H0} = 19.8 s_{0m}^{-0.5} \exp(-7.08/H_0) \qquad \text{for } T_0 \geq T_0^*$$
$$f_{H0} = 0.05\, H_0 T_0 + 10.5 \qquad \text{for } T_0 < T_0^* \qquad 6.9$$

where $H_0 = H_{m0}/\Delta D_{n50}$, $T_0 = (g/D_{n50})^{0.5} T_{m0,1}$,
and $T_0^* = \{19.8\, s_{0m}^{-0.5} \exp(-7.08/H_0) - 10.5\}/(0.05\, H_0)$.

The berm level d_h is also taken into account as an influence factor, d_h^*. Note that the berm depth is positive if the berm level is below SWL, and therefore, for berm breakwaters often negative. Note also that this influence factor is different than for a bermed slope, see Section 5.3.4. This influence factor is described by:

$$d_h^* = (3H_{m0} - d_h)/(3H_{m0} + R_c) \qquad \text{for } d_h < 3H_{m0}$$
$$d_h^* = 0 \qquad \text{for } d_h \geq 3H_{m0} \qquad 6.10$$

The final overtopping formula then takes into account the influence factor on recession, f_{H0}, the influence factor of the berm level, d_h^*, the geometrical parameters R_c, B and G_c, the wave conditions H_{m0} and the mean period $T_{m0,1}$. It means that the wave overtopping is described by a spectral mean period, not by $T_{m-1,0}$.

$$\frac{q}{(gH_{m0}^3)^{0.5}} = 1.79 \times 10^{-5} \left(f_{H0}^{1.34} + 9.22 \right) s_{op}^{-2.52} e^{\left(-5.63 \left(\frac{R_c}{H_{m0}} \right)^{0.92} - 0.61 \left(\frac{G_c}{H_{m0}} \right)^{1.39} - 0.55 h_{b*}^{1.48} \left(\frac{B}{H_{m0}} \right)^{1.39} \right)} \quad 6.11$$

Equation 6.11 is only valid for a lower slope of 1:1.25 and an upper slope of 1:1.25. For other slopes one has to reshape the slope to a slope of 1:1.25, keeping the volume of material the same and adjusting the berm width B and for the upper slope also the crest width G_c. Note also that in Equation 6.11 the peak wave period T_p has to be used to calculate s_{op}, where the mean period $T_{m0,1}$ has to be used in Equation 6.9.

Although no tests were performed on the non-reshaping Icelandic berm breakwaters (see Fig. 6.7), a number of tests were performed on non-reshaping structures by keeping the material in place with a steel net. The difference may be that Icelandic berm breakwaters show a little less overtopping, due to the presence of larger rock and, therefore, more permeability. The tests showed that Equation 6.11 is also valid for non-reshaping berm breakwaters, if the reshaping factor $f_{H0} = 0$.

6.3.5 Effect of wave walls

Most breakwaters have a wave wall, capping wall or crest unit on the crest, simply to end the armour layer in a good way and to create access to the breakwater. For design it is advised not to design a wave wall much higher than the armour crest, for the simple reason that wave forces on the wall will increase drastically if directly attacked by waves and not hidden behind the armour crest. For rubble mound slopes as a shore protection, design waves might be a little lower than for breakwaters and a wave wall might be one of the solutions to reduce wave overtopping. Nevertheless, one should realise the increase in wave forces if designing a wave wall significantly above the armour crest.

Equations 6.5 and 6.6 for a simple rubble mound slope includes a berm of $3D_n$ wide and a wave wall at the same level as the armour crest: $A_c = R_c$. A little lower wave wall will hardly give larger overtopping, but no wave wall at all would certainly increase overtopping. Part of the overtopping waves will then penetrate through the crest armour. No formula are present to cope with such a situation, unless the use of the Neural Network prediction method (Section 4.4).

Various researchers have investigated wave walls higher than the armour crest. None of them compared their results with a graph like Fig. 6.6 for simple rubble mound slopes. During the writing of this manual some of the published equations were plotted in Fig. 6.6 and most curves fell within the scatter of the data. Data with a wider crest gave significantly lower overtopping, but that was due to the wider crest, not the higher wave wall. In essence the message is: use the height of the wave wall R_c and not the height of the armoured crest A_c in Equations 6.5 and 6.6 if the wall is higher than the crest. For a wave wall lower than the crest armour the height of this crest armour should be used. The Neural Network prediction might be able to give more precise predictions.

6.3.6 Scale and model effect corrections

Results of the recent CLASH project suggested significant differences between field and model results on wave overtopping. This has been verified for different sloping rubble structures. Results of the comparisons in this project have led to a scaling procedure which is mainly dependent on the roughness of the structure γ_f [–]; the seaward slope cot α of the structure [–]; the mean overtopping discharge, up-scaled to prototype, q_{ss} [m³/s/m]; and whether wind is considered or not.

Data from the field are naturally scarce, and hence the method can only be regarded as tentative. It is furthermore only relevant if mean overtopping rates are lower than 1.0 l/s/m but may include significant adjustment factors below these rates. Due to the inherent uncertainties, the proposed approach tries to be conservative. It has however been applied to pilot cases in CLASH and has proved good corrections with these model data.

The adjustment factor fq for model and scale effects can be determined as follows:*

$$f_q = \begin{cases} 1.0 & \text{for } q_{ss} \geq 10^{-3} \text{ m}^3/\text{s/m} \\ \min\{(-\log q_{ss} - 2)^3; f_{q,\max}\} & \text{for } q_{ss} \geq 10^{-3} \text{ m}^3/\text{s/m} \end{cases} \quad 6.12$$

where $f_{q,\max}$ is an upper bound to the adjustment factor f_q and can be calculated as follows:

$$f_{q,\max} = \begin{cases} f_{q,r} & \text{for } \gamma_f \leq 0.7 \\ 5 \cdot \gamma_f \cdot (1 - f_{q,r}) + 4.5 \cdot (f_{q,r} - 1) + 1 & \text{for } 0.7 < \gamma_f \leq 0.9 \\ 1.0 & \text{for } \gamma_f > 0.9 \end{cases} \quad 6.13$$

and $f_{q,r}$ is the adjustment factor for rough slopes which is mainly dependent on the slope of the structure and whether wind needs to be included or not.

$$f_{q,r} = \begin{cases} 1.0 & \text{for } \cot\alpha \leq 0.6 \\ f_w \cdot (8.5 \cdot \cot\alpha - 4.0) & \text{for } 0.6 < \cot\alpha \leq 4.0 \\ f_w \cdot 30 & \text{for } \cot\alpha > 4.0 \end{cases} \quad 6.14$$

in which f_w accounts for the presence of wind and is set to $f_w = 1.0$ if there is wind and $f_w = 0.67$ if there is no wind.

This set of equations include the case of smooth dikes which will – due to $\gamma_f = 0.9$ in this case – always lead to an adjustment factor of $f_q = 1.0$. In case of a very rough 1:4 slope with wind $f_{q,\max} = f_{qr} = 30.0$ which is the maximum the factor can get to (but only if the mean overtopping rates gets below $q_{ss} = 10^{-5}$ m³/s/m). The latter case and a steep rough slope is illustrated in Fig. 6.8.

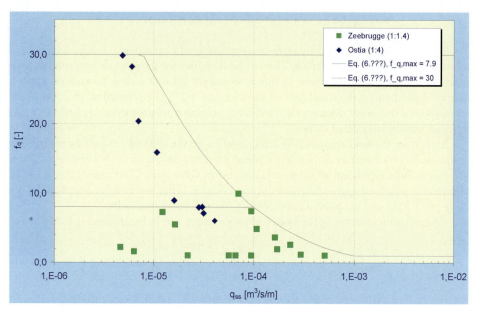

Fig. 6.10: Proposed adjustment factor applied to data from two field sites (Zeebrugge 1:1.4 rubble mound breakwater, and Ostia 1:4 rubble slope)

6.4 Overtopping volumes per wave

Wave overtopping is a dynamic and irregular process and the mean overtopping discharge, q, does not cover this aspect. But by knowing the storm duration, t, and the number of overtopping waves in that period, N_{ow}, it is easy to describe this irregular and dynamic overtopping, if the overtopping discharge, q, is known. Each overtopping wave gives a certain overtopping volume of water, V. The general distribution of overtopping volumes for coastal structures has been described in Section 4.2.2.

As with many equations in this manual, the two-parameter Weibull distribution describes the behaviour quite well. This equation has a shape parameter, b, and a scale parameter, a. For smooth sloping structures an average value of b = 0.75 was found to indicate the distribution of overtopping volumes (see Section 5.4). The same average value will be used for rubble mound structures, which makes smooth and rubble mound structures easily comparable. The exceedance probability, P_V, of an overtopping volume per wave is then similar to Equations 4.2 and 4.3.

$$P_V = P(\underline{V} \leq V) = 1 - \exp\left[-\left(\frac{V}{a}\right)^{0.75}\right] \qquad 6.15$$

with:

$$a = 0.84 \cdot T_m \cdot \frac{q}{P_{ov}} = 0.84 \cdot T_m \cdot q \cdot N_w / N_{ow} = 0.84 \cdot q \cdot t / N_{ow} \qquad 6.16$$

Equation 6.16 shows that the scale parameter depends on the overtopping discharge, but also on the mean period and probability of overtopping, or which is similar, on the storm duration and the actual number of overtopping waves.

The probability of wave overtopping for rubble mound structures has been described in Section 6.2, Fig. 6.4 and Equation 6.4.

Equations for calculating the overtopping volume per wave for a given probability of exceedance, is given by Equation 5.34. The maximum overtopping during a certain event is fairly uncertain, as most maxima, but depends on the duration of the event. In a 6 hours period one may expect a larger maximum than only during 15 minutes. The maximum during an event can be calculated by Equation 5.35.

6.5 Overtopping velocities and spatial distribution

The hydraulic behaviour of waves on rubble mound slopes and on smooth slopes like dikes, is generally based on similar formulae, as clearly shown in this chapter. This is different, however, for overtopping velocities and spatial distribution of the overtopping water. A dike or sloping impermeable seawall generally has an impermeable and more or less horizontal crest. Up-rushing and overtopping waves flow over the crest and each overtopping wave can be described by a maximum velocity and flow depth, see Section 5.5. These velocities and flow depths form the description of the hydraulic loads on crest and inner slope and are part of the failure mechanism "failure or erosion of inner slopes by wave overtopping".

This is different for rubble mound slopes or breakwaters where wave energy is dissipated in the rough and permeable crest and where often overtopping water falls over a crest wall onto a crest road or even on the rear slope of a breakwater. A lot of overtopping water travels over the crest and through the air before it hits something else.

Only recently in CLASH and a few other projects at Aalborg University attention has been paid to the spatial distribution of overtopping water at breakwaters with a crest wall (LYKKE ANDERSEN and BURCHARTH, 2006). The spatial distribution was measured by various trays behind the crest wall. Fig. 6.11 gives different cross-sections with a set-up of three arrays. Up to six arrays have been used. The spatial distribution depends on the level with

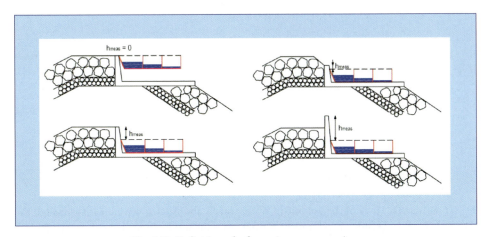

Fig. 6.11: Definition of y for various cross-sections

respect to the rear side of the crest wall and the distance from this rear wall, see Fig. 6.12. The coordinate system (x, y) starts at the rear side and at the top of the crest wall, with the positive y-axis downward.

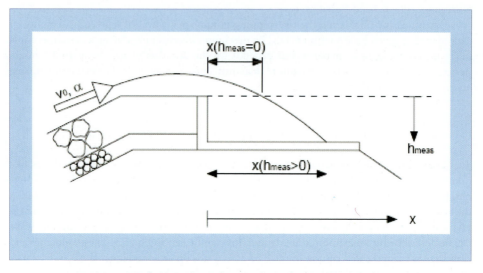

Fig. 6.12: Definition of x- and y-coordinate for spatial distribution

The exceedance probability F of the travel distance is defined as the volume of overtopping water passing a given x- and y-coordinate, divided by the total overtopping volume. The probability, therefore, lies between 0 and 1, with 1 at the crest wall. The spatial distribution can be described with the following equations, which have slightly been rewritten and modified with respect to the original formulae by LYKKE ANDERSEN and BURCHARTH (2006). The probability F at a certain location can be described by:

$$F(x,y) = \exp\left(\frac{-1.3}{H_{m0}} \cdot \left\{\max\left(\frac{x}{\cos\beta} - 2.7 y s_{op}^{0.15}, 0\right)\right\}\right) \qquad 6.17$$

Equation 6.17 can be rewritten to calculate the travel distance x directly (at a certain level y) by rewriting the above equation:

$$\frac{x}{\cos\beta} = -0.77 H_{m0} \ln(F) + 2.7 y s_{op}^{0.15} \qquad 6.18$$

Suppose cosβ = 0, then we get:
  F = 1   x = 0
  F = 0.1   x = 1.77 H_s
  F = 0.01   x = 3.55 H_s

It means that 10 % of the volume of water travels almost two wave heights through the air and 1 % of the volume travels more than 3.5 times the wave height. These percentages will be higher if y ≠ 0, which is often the case with a crest unit.

The validity of Equations 6.17 and 6.18 is for rubble mound slopes of approximately 1:2 and for angles of wave attack between $0° \leq |\beta| < 45°$. It should be noted that the equation is valid for the spatial distribution of the water through the air behind the crest wall. All water falling on the basement of the crest unit will of course travel on and will fall into the water behind and/or on the slope behind.

6.6 Overtopping of shingle beaches

Shingle beaches differ from the armoured slopes principally in the size of the beach material, and hence its mobility. The typical stone size is sufficiently small to permit significant changes of beach profile, even under relatively low levels of wave attack. A shingle beach may be expected to adjust its profile to the incident wave conditions, provided that sufficient beach material is available. Run-up or overtopping levels on a shingle beach are therefore calculated without reference to any initial slope.

The equilibrium profile of shingle beaches under (temporary constant) wave conditions is described by VAN DER MEER (1988). The most important profile parameter for run-up and overtopping is the crest height above SWL, h_c. For shingle with $D_{n50} < 0.1$ m this crest height is only a function of the wave height and wave steepness. Note that the mean wave period is used, not the spectral wave period $T_{m-1,0}$.

$$h_c/H_{m0} = 0.3 \, s_{om}^{-0.5} \qquad 6.19$$

Only the highest waves will overtop the beach crest and most of this water will percolate through the material behind the beach crest. Equation 6.19 gives a run-up or overtopping level which is more or less close to $Ru_{2\%}$.

6.7 Uncertainties

Since wave overtopping formulae are principally identical to the ones for sea dikes, uncertainties of the models proposed in this chapter should be dealt with in the same way as those proposed in section 5.8 already.

It should however be noted that some of the uncertainties of the relevant parameters might change. For rubble mound structures the crest height is about 30 % more uncertain than for smooth dikes and will result in about 0.08 m. Furthermore, the slope uncertainty increases by about 40 % to 2.8 %. All uncertainties related to waves and water levels will remain as discussed within section 5.8.

The minor changes in these uncertainties will not affect the lines as shown in Fig. 5.43. Hence, the same proposal accounting for uncertainties as already given in Section 5.8 is applied here.

Again, it should be noted that only uncertainties for mean wave overtopping rates are considered here. Other methods as discussed in this chapter were disregarded but can be dealt with using the principal procedure as discussed in Section 1.5.4.

7. Vertical and step seawalls

7.1 Introduction

This chapter presents guidance for the assessment of overtopping and post-overtopping processes at vertical and steep-fronted coastal structures such as caisson and blockwork breakwaters and vertical seawalls (Fig. 7.1, Fig. 7.2). Also included are composite vertical wall structures (where the emergent part of the structure is vertical, fronted by a modest berm) and vertical structures which include a recurve/bull-nose/parapet/wave return wall as the upper part of the defence.

Large vertical breakwaters (Fig. 7.1) are almost universally formed of sand-filled concrete caissons usually resting on a small rock mound. Such caisson breakwaters may reach depths greater than 100 m, under which conditions no wave breaking at all at the wall would be expected. Conversely, older breakwaters may, out of necessity, have been constructed in shallower water or indeed, built directly on natural rock "skerries". As such, these structures may find themselves exposed to breaking wave, or "impulsive" conditions when the water depth in front of them is sufficiently low. Urban seawalls (*e.g.* Fig. 7.2) are almost universally fronted by shallow water, and are likely to be exposed to breaking or broken wave conditions, especially in areas of significant tidal range.

Fig. 7.1: Examples of vertical breakwaters: (left) modern concrete caisson and (right) older structure constructed from concrete blocks

Fig. 7.2: Examples of vertical seawalls: (left) modern concrete wall and (right) older stone blockwork wall

There are three principal sources of guidance on this topic preceding this manual; in the UK, the Environment Agency "Overtopping of Seawalls: Design and Assessment Manual" (EA/BESLEY, 1999); in the U.S.A., the US Army Corps of Engineers' "Coastal Engineering Manual" (CEM/BURCHARTH & HUGHES, 2002); in Japan, Goda's design charts (*e.g.* GODA, 2000). The guidance presented in this chapter builds upon that of EA/BESLEY (1999), with adjustments to many formulae based upon further testing since 1999.

For those familiar with EA/BESLEY (1999), the principal changes/additions are

- new guidance on prediction of mean and wave-by-wave overtopping to oblique wave attack under impulsive conditions (Section 7.3.4);
- extension of method for mean overtopping to account for steep (but not vertical) "battered" walls (Section 7.3.2);
- new guidance on mean overtopping under conditions when all waves break before reaching the wall (part of Section 7.3.1);
- new guidance on reduction in mean overtopping discharge due to wave return walls/parapets / recurves (Section 7.3.5);
- new guidance on "post-overtopping" processes, specifically; velocity of "throw"; landward spatial extent of overtopping, and effect of wind (Section 7.3.6)
- inclusion of summary of new evidence on scale effects for laboratory study of overtopping at vertical and steep walls (Section 7.3.7).
- minor adjustments to recommended approach for distinguishing impulsive/non-impulsive conditions (Section 7.2);
- minor adjustments to formulae for mean overtopping under impulsive conditions due to the availability of additional data, from e.g. the CLASH database (Section 7.3.1).
- all formulae are now given in terms of wave period $T_{m-1,0}$ resulting in an adjusted definition of the $h*$ and $d*$ parameters (Sections 7.2.2 and 7.2.3 respectively) in order to maintain comparability with earlier work.
- in line with convergence on the $T_{m-1,0}$ measure, formulae using wave steepness s_{op} have been adjusted to use the new preferred measure $s_{m-1,0}$ (Section 7.3.1);
- all formulae are now given explicitly in terms of basic wave and structural parameters without recourse to intermediate definitions of dimensionless overtopping discharge and freeboard parameters specific to impulsive conditions.

This chapter follows approximately the same sequence as the preceding two chapters, though certain differences should be noted. In particular, run-up is not addressed, as it is not a measure of physical importance for this class of structure – indeed it is not well-defined for cases when the wave breaks, nearly-breaks or is broken when it reaches the structure, under which conditions an up-rushing jet of water is thrown upwards.

The qualitative form of the physical processes occurring when the waves reach the wall are described in Section 7.2. Distinctions drawn between different wave/structure "regimes" are reflected in the guidance for assessment of mean overtopping discharges given in Section 7.3. The basic assessment tools are presented for plain vertical walls (Section 7.3.1), followed by subsections giving advice on how these basic tools should be adjusted to account for other commonly-occurring configurations; battered walls (Section 7.3.2); vertically composite walls (Section 7.3.3); the effect of oblique wave attack (Section 7.3.4); the effect of recurve/wave-return walls (Section 7.3.5). Scale and model effects are reviewed in Section 7.3.7. Methods to assess individual "wave by wave" overtopping volumes are presented in Section 7.4. The current knowledge and advice on post-overtopping processes including velocities, spatial distributions and post-overtopping loadings are reviewed in Section 7.5.

Principal calculation procedures are summarised in Table 7.1.

Table 7.1: Summary of principal calculation procedures for vertical structures

	Deterministic design	Probabilistic design
Discrimination – impulsive / non-impulsive regime		
plain vertical walls		Eq. 7.1
vertical composite walls		Eq. 7.2
Plain vertical walls		
non-impulsive conditions	Eq. 7.4	Eq. 7.3
impulsive conditions	Eq. 7.6	Eq. 7.5
broken wave conditions (submerged toe)	Eq. 7.8	Eq. 7.7
broken wave conditions (emergent toe)	Eq. 7.10	Eq. 7.9
Battered walls	Eq. 7.11	Eq. 7.11
Composite vertical walls	Eq. 7.13	Eq. 7.12
Oblique wave attack		
non-impulsive conditions		Eq. 7.14 & 7.15
impulsive conditions	Eq. 7.17	Eq. 7.18
Vertical walls with wave return wall / parapet		Eqs. 7.18, 7.19 & Fig. 7.20
Effect of wind		Eq. 7.20 & 7.21
Percentage of overtopping waves		Eq. 7.22 / 7.23
with oblique waves		Eqs. 7.29 & 7.30
Individual overtopping volumes		Eqs. 7.24 to 7.28
with oblique waves		Table 7.2
Overtopping velocities		Eq. 7.31
Spatial extent of overtopping		Fig. 7.25
Downfall pressures due to overtopped discharge		Eq. 7.32

7.2 Wave processes at walls

7.2.1 Overview

In assessing overtopping on sloping structures, it is necessary to distinguish whether waves are in the "plunging" or "surging" regime (Section 5.3.1). Similarly, for assessment of overtopping at steep-fronted and vertical structures the regime of the wave/structure interaction must be identified first, with quite distinct overtopping responses expected for each regime.

On steep walls (vertical, battered or composite), "non-impulsive" or "pulsating" conditions occur when waves are relatively small in relation to the local water depth, and of lower wave steepnesses. These waves are not critically influenced by the structure toe or approach slope. Overtopping waves run up and over the wall giving rise to (fairly) smoothly-varying loads and "green water" overtopping (Fig. 7.3).

In contrast, "impulsive" conditions (Fig. 7.4) occur on vertical or steep walls when waves are larger in relation to local water depths, perhaps shoaling up over the approach bathymetry or structure toe itself. Under these conditions, some waves will break violently against the wall with (short-duration) forces reaching 1õ– 40 times greater than for non-impulsive conditions. Overtopping discharge under these conditions is characterised by a "violent" uprushing jet of (probably highly aerated) water.

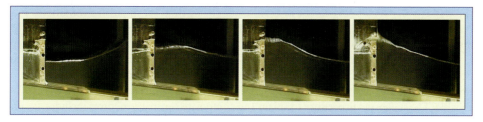

Fig. 7.3: A non-impulsive (pulsating) wave condition at a vertical wall, resulting in non-impulsive (or "green water") overtopping

Fig. 7.4: An impulsive (breaking) wave at a vertical wall, resulting in an impulsive (violent) overtopping condition

Lying in a narrow band between non-impulsive and impulsive conditions are "near-breaking" conditions where the overtopping is characterised by suddenness and a high-speed, near vertical up-rushing jet (like impulsive conditions) but where the wave has not quite broken onto the structure and so has not entrained the amount of air associated with fully impulsive conditions. This "near-breaking" condition is also known as the "flip through" condition. This conditions gives overtopping in line with impulsive (breaking) conditions and are thus not treated separately.

Many seawalls are constructed at the back of a beach such that breaking waves never reach the seawall, at least not during frequent events where overtopping is of primary importance. For these conditions, particularly for typical shallow beach slopes of less than (say) 1:30, design wave conditions may be given by waves which start breaking (possibly quite some distance) seaward of the wall. These "broken waves" arrive at the wall as a highly-aerated mass of water (Fig. 7.5), giving rise to loadings which show the sort of short-duration peak seen under impulsive conditions (as the leading edge of the mass of water arrives at the wall) but smaller in magnitude due to the high level of aeration. For cases where the depth at

Fig. 7.5: A broken wave at a vertical wall, resulting in a broken wave overtopping condition

the wall $h_s > 0$, overtopping can be assessed using the method for impulsive conditions. For conditions where the toe of the wall is emergent ($h_s \leq 0$), these methods can no longer be applied and an alternative is required (Section 7.3.1).

In order to proceed with assessment of overtopping, it is therefore necessary first to determine which is the dominant overtopping regime (impulsive or non-impulsive) for a given structure and design sea state. No single method gives a discriminator which is 100 % reliable. The suggested procedure for plain and composite vertical structures includes a transition zone in which there is significant uncertainty in the prediction of dominant overtopping regime and thus a "worst-case" is taken.

7.2.2 Overtopping regime discrimination – plain vertical walls

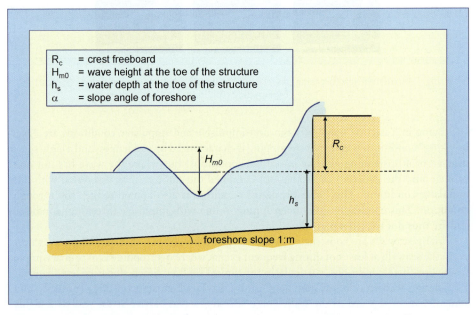

Fig. 7.6: Definition sketch for assessment of overtopping at plain vertical walls

This method is for distinguishing between impulsive and non-impulsive conditions at a vertical wall where the toe of the wall is submerged ($h_s > 0$; Fig. 7.6). When the toe of the wall is emergent ($h_s < 0$) only broken waves reach the wall.

For submerged toes ($h_s > 0$), a wave breaking or "impulsiveness" parameter, $h*$ is defined based on depth at the toe of the wall, h_s, and incident wave conditions inshore:

$$h_* = 1.35 \frac{h_s}{H_{m0}} \frac{2\pi h_s}{g T_{m-1,0}^2} \qquad 7.1$$

Non-impulsive (pulsating) conditions dominate at the wall when $h* > 0.3$, and impulsive conditions occur when $h* < 0.2$. The transition between conditions for which the overtopping response is dominated by breaking and non-breaking waves lies over $0.2 \leq h* \leq 0.3$. In this region, overtopping should be predicted for both non-impulsive and impulsive conditions, and the larger value assumed.

7.2.3 Overtopping regime discrimination – composite vertical walls

For vertical composite walls where a berm or significant toe is present in front of the wall, an adjusted version of the method for plain vertical walls should be used. A modified "impulsiveness" parameter, $d*$, is defined in a similar manner to the $h*$ parameter (for plain vertical walls, Section 7.2.2);

$$d_* = 1.35 \frac{d}{H_{m0}} \frac{2\pi h_s}{g T_{m-1,0}^2} \qquad 7.2$$

with parameters defined according to Fig. 7.7.

Non-impulsive conditions dominate at the wall when $d* > 0.3$, and impulsive conditions occur when $d* < 0.2$. The transition between conditions for which the overtopping response is dominated by breaking and non-breaking waves lies over $0.2 \leq d* \leq 0.3$. In this region, overtopping should be predicted for both non-impulsive and impulsive conditions, and the larger value assumed.

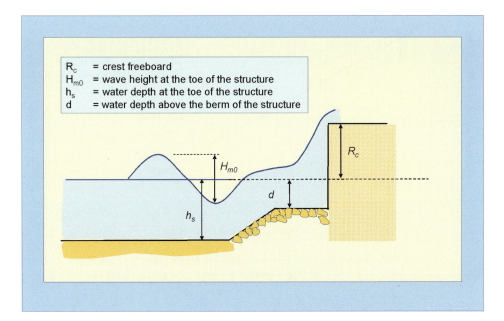

Fig. 7.7: Definition sketch for assessment of overtopping at composite vertical walls

7.3 Mean overtopping discharges for vertical and battered walls

7.3.1 Plain vertical walls

For simple vertical breakwaters under the following equations should be used:

Probabilistic design, non-impulsive conditions ($h* > 0.3$): The mean prediction should be used for probabilistic design, or for comparison with measurements (Equation 7.3). The coefficient of 2.6 for the mean prediction has an associated standard deviation of $\sigma = 0.8$.

$$\frac{q}{\sqrt{gH_{m0}^3}} = 0.04\exp\left(-2.6\frac{R_c}{H_{m0}}\right) \qquad \text{valid for } 0.1 < R_c/H_{m0} < 3.5 \qquad 7.3$$

Deterministic design or safety assessment, non-impulsive conditions ($h* > 0.3$): For deterministic design or safety assessment, the following equation incorporates a factor of safety of one standard deviation above the mean prediction:

$$\frac{q}{\sqrt{gH_{m0}^3}} = 0.04\exp\left(-1.8\frac{R_c}{H_{m0}}\right) \qquad \text{valid for } 0.1 < R_c/H_{m0} < 3.5 \qquad 7.4$$

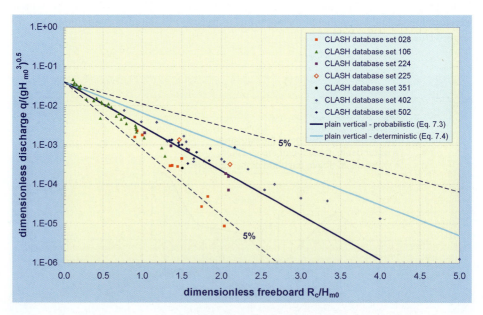

Fig. 7.8: Mean overtopping at a plain vertical wall under non-impulsive conditions (Equations 7.3 and 7.4)

Zero Freeboard: For a vertical wall under non-impulsive conditions Equation 7.5 should be used for probabilistic design and for prediction and comparison of measurements (Fig. 5.13) SMID (2001).

$$\frac{q}{\sqrt{gH_{m0}^3}} = 0.062 \pm 0.0062 \qquad \text{valid for } R_c/H_{m0} = 0 \qquad 7.5$$

For deterministic design or safety assessment it is recommended to increase the average overtopping discharge in Equation 7.5 by one standard deviation.

No data are available for impulsive overtopping at zero freeboard at vertical walls.

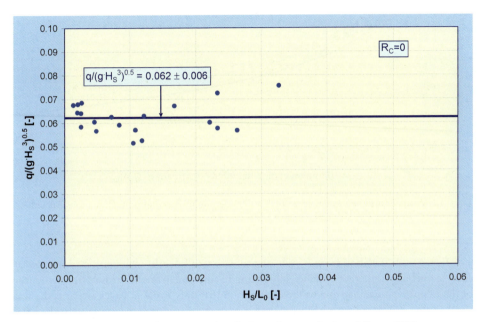

Fig. 7.9: Dimensionless overtopping discharge for zero freeboard (SMID, 2001)

Probabilistic design, impulsive conditions ($h* \leq 0.2$): The mean prediction should be used for probabilistic design, or for comparison with measurements (Equation 7.6). The scatter *in the logarithm* of the data about the mean prediction is characterised by a standard deviation of c. 0.37 (*i.e.* c. 68 % of predictions lie within a range of ×/÷ 2.3).

$$\frac{q}{h_*^2 \sqrt{gh_s^3}} = 1.5 \times 10^{-4} \left(h_* \frac{R_c}{H_{m0}} \right)^{-3.1} \qquad \text{valid over } 0.03 < h_* \frac{R_c}{H_{m0}} < 1.0 \qquad 7.6$$

Deterministic design or safety assessment, impulsive conditions ($h* \leq 0.2$): For deterministic design or safety assessment, the following equation incorporates a factor of safety of one standard deviation above the mean prediction:

$$\frac{q}{h_*^2 \sqrt{gh_s^3}} = 2.8 \times 10^{-4} \left(h_* \frac{R_c}{H_{m0}} \right)^{-3.1} \qquad \text{valid over } 0.03 < h_* \frac{R_c}{H_{m0}} < 1.0 \qquad 7.7$$

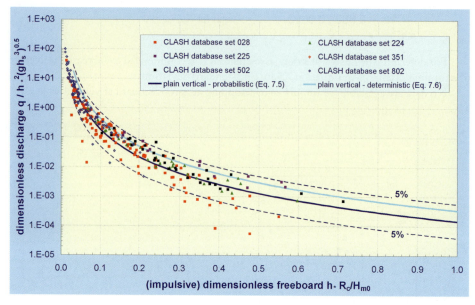

Fig. 7.10: Mean overtopping at a plain vertical wall under impulsive conditions (Equations 7.6 and 7.7)

For $R_b < 0.02$ arising from h_s reducing to very small depths (as opposed to from small relative freeboards) there is evidence supporting an adjustment downwards of the predictions of the impulsive formulae due to the observation that only broken waves arrive at the wall (BRUCE et al., 2003). For probabilistic design or comparison with measurements, the mean prediction should be used (Equation 7.8). The scatter in the logarithm of the data about the mean prediction is characterised by a standard deviation of c. 0.15 (i.e. c. 68 % of predictions lie within a range of $\times / \div 1.4$).

$$\frac{q}{h_*^2 \sqrt{gh_s^3}} = 2.7 \times 10^{-4} \left(h_* \frac{R_c}{H_{m0}} \right)^{-2.7} \text{ valid for } h_* \frac{R_c}{H_{m0}} < 0.02; \text{ broken waves} \qquad 7.8$$

For **deterministic design or safety assessment,** the following equation incorporates a factor of safety of one standard deviation (in the multiplier) above the mean prediction:

$$\frac{q}{h_*^2 \sqrt{gh_s^3}} = 3.8 \times 10^{-4} \left(h_* \frac{R_c}{H_{m0}} \right)^{-2.7} \text{ valid for } h_* \frac{R_c}{H_{m0}} < 0.02; \text{ broken waves} \qquad 7.9$$

For $0.02 < h_* R_c / H_{m0} < 0.03$, there appears to be a transition between Equation 7.7 (for "normal" impulsive conditions) and Equation 7.8 (for conditions with only broken waves). There is however insufficient data upon which to base a firm recommendation in this range. It is suggested that Equation 7.7 is used down to $h_* R_c / H_{m0} = 0.02$ unless it is clear that only broken waves will arrive at the wall, in which case Equation 7.8 could be used. Formulae for these low $h_* R_c / H_{m0}$ conditions are shown in Fig. 7.11.

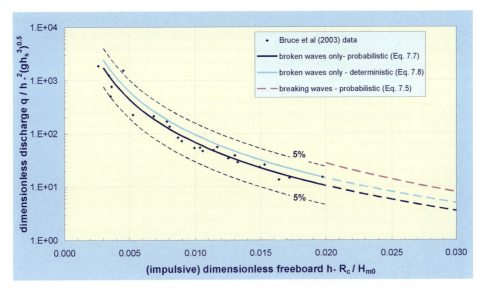

Fig. 7.11: Mean overtopping discharge for lowest $h* R_c / H_{m0}$ (for broken waves only arriving at wall) with submerged toe ($h_s > 0$). For $0.02 < h* R_c / H_{m0} < 0.03$, overtopping response is ill-defined – lines for both impulsive conditions (extrapolated to lower $h* R_c / H_{m0}$) and broken wave only conditions (extrapolated to higher $h* R_c / H_{m0}$) are shown as dashed lines over this region

Data for configurations where the toe of the wall is emergent (*i.e.* at or above still water level, $h_s \leq 0$) is limited. The only available study suggests an adaptation of a prediction equation for plunging waves on a smooth slope may be used, but particular caution should be exercised in any extrapolation beyond the parameter ranges of the study, which only used a relatively steep ($m = 10$) foreshore slope.

For **probabilistic design or comparison with measurements,** the mean prediction should be used (Equation 7.10) should be used. The standard deviation associated with the exponent coefficient (–2.16) is *c.* 0.21.

$$\frac{q}{\sqrt{gH_{m0,deep}^3}} \cdot \sqrt{m s_{m-1,0}} = 0.043 \exp\left(-2.16 m s_{m-1,0}^{0.33} \frac{R_c}{H_{m0,deep}}\right) \qquad 7.10$$

valid for $2.0 < m s_{m-1,0}^{0.33} \frac{R_c}{H_{m0,deep}} < 5.0$; $0.55 \leq R_c/H_{m0,deep}$

$s_{m-1,0} \geq 0.025$; Note – data only available for $m = 10$ (*i.e.* 1:10 foreshore slope)

For **deterministic design or safety assessment,** Equation 7.11 incorporates a factor of safety of one standard deviation (in the exponent) above the mean prediction.

$$\frac{q}{\sqrt{gH_{m0,deep}^3}} \cdot \sqrt{m s_{m-1,0}} = 0.043 \exp\left(-1.95 m s_{m-1,0}^{0.33} \frac{R_c}{H_{m0,deep}}\right) \qquad 7.11$$

valid for $2.0 < m s_{m-1,0}^{0.33} \frac{R_c}{H_{m0,deep}} < 5.0$; $0.55 \leq R_c/H_{m0,deep} \leq 1.6$;

$s_{m-1,0} \geq 0.025$; NB – data only available for $m = 10$ (*i.e.* 1:10 foreshore slope)

Equations 7.10 and 7.11 for overtopping under emergent toe conditions are illustrated in Fig. 7.12. It should be noted that this formula is based upon a limited dataset of small-scale tests with 1:10 foreshore only and should not be extrapolated beyond the ranges tested (foreshore slope 1:m = 0.1; $s_{op} \geq 0.025$; $0.55 \leq R_c/H_{m0,deep} \leq 1.6$).

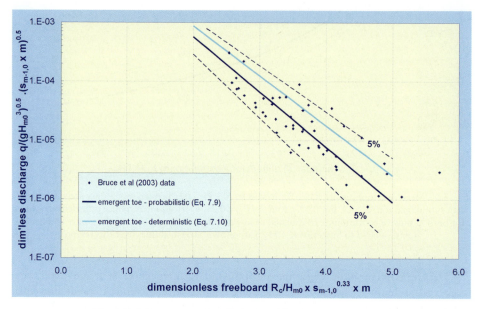

Fig. 7.12: Mean overtopping discharge with emergent toe ($h_s < 0$)

7.3.2 Battered walls

Near-vertical walls with 10:1 and 5:1 batters are found commonly for older UK seawalls and breakwaters (*e.g.* Fig. 7.13).

Mean overtopping discharges for battered walls under impulsive conditions are slightly in excess of those for a vertical wall over a wide range of dimensionless freeboards. Multiplying factors are given in Equation 7.12 (plotted in Fig. 7.14).

10:1 battered wall: $q_{10:1\ batter} = q_{vertical} \times 1.3$

5:1 battered wall: $q_{5:1\ batter} = q_{vertical} \times 1.9$

7.12

where $q_{vertical}$ is arrived at from Equation 7.6 (for probabilistic design) or Equation 7.7 (for deterministic design). The uncertainty in the final estimated overtopping discharge can be estimated as per the plain vertical cases.

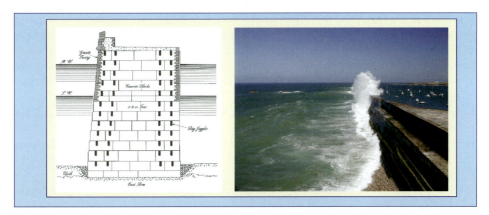

Fig. 7.13: Battered walls: typical cross-section (left), and Admiralty Breakwater, Alderney Channel Islands (right, courtesy G. MÜLLER)

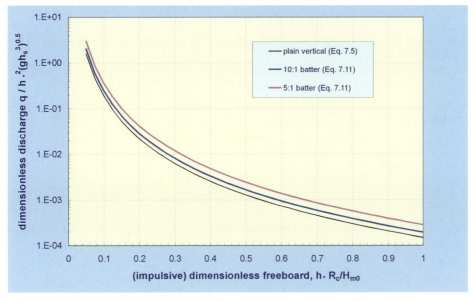

Fig. 7.14: Overtopping for a 10:1 and 5:1 battered walls

No dataset is available to indicate an appropriate adjustment under non-impulsive conditions. Given that these battered structures are generally older structures in shallower water, it is likely that impulsive conditions are possible at most, and will form the design case.

7.3.3 Composite vertical walls

It is well-established that a relatively small toe berm can change wave breaking characteristics, thus substantially altering the type and magnitude of wave loadings (e.g. (OUMERACI et al., 2001). Many vertical seawall walls may be fronted by rock mounds with the intention of protecting the toe of the wall from scour. The toe configuration can vary considerably, potentially modifying the overtopping behaviour of the structure. Three types of mound can be identified

1. Small toe mounds which have an insignificant effect on the waves approaching the wall – here the toe may be ignored and calculations proceed as for simple vertical (or battered) walls.
2. Moderate mounds, which significantly affect wave breaking conditions, but are still below water level. Here a modified approach is required.
3. Emergent mounds in which the crest of the armour protrudes above still water level. Prediction methods for these structures may be adapted from those for crown walls on a rubble mound (Section 6.3.5).

For assessment of mean overtopping discharge at a composite vertical seawall or breakwater, the overtopping regime (impulsive/non-impulsive) must be determined – see Section 7.2.3.

When non-impulsive conditions prevail, overtopping can be predicted by the standard method given previously for non-impulsive conditions at plain vertical structures, Equation 7.3.

For conditions determined to be impulsive, a modified version of the impulsive prediction method for plain vertical walls is recommended, accounting for the presence of the mound by use of d and d_*.

For **probabilistic design or comparison with measurements**, the mean prediction (Equation 7.13) should be used. The scatter *in the logarithm* of the data about the mean prediction is characterised by a standard deviation of $c.$ 0.28 (*i.e. c.* 68 % of predictions lie within a range of ×/÷ 1.9).

$$\frac{q}{d_*^2 \sqrt{gh_s^3}} = 4.1 \times 10^{-4} \left(d_* \frac{R_c}{H_{m0}} \right)^{-2.9}$$
7.13

valid for $0.05 < d_* \frac{R_c}{H_{m0}} < 1.0$ and $h_* < 0.3$

For **deterministic design or safety assessment**, Equation 7.14 incorporates a factor of safety of one standard deviation (in the constant multiplier) above the mean prediction.

$$\frac{q}{d_*^2 \sqrt{gh_s^3}} = 7.8 \times 10^{-4} \left(d_* \frac{R_c}{H_{m0}} \right)^{-2.6}$$
7.14

valid for $0.05 < d_* \frac{R_c}{H_{m0}} < 1.0$ and $h_* < 0.3$

7.3.4 Effect of oblique waves

Seawalls and breakwaters seldom align perfectly with incoming waves. The assessment methods presented thus far are only valid for shore-normal wave attack. In this subsection, advice on how the methods for shore-normal wave attack (obliquity β = 0°) should be adjusted for oblique wave attack.

This chapter extends the existing design guidance for impulsive wave attack from perpendicular to oblique wave attack. As for zero obliquity, overtopping response depends

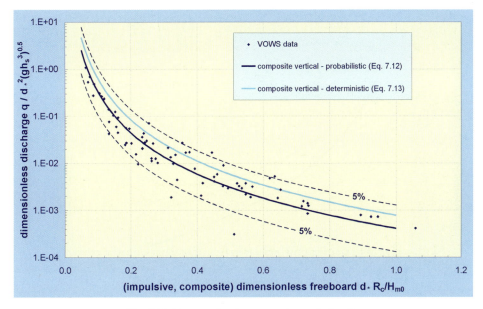

Fig. 7.15: Overtopping for composite vertical walls

critically upon the physical form (or "regime") of the wave/wall interaction – non-impulsive; impulsive or broken. As such, the first step is to use the methods given in Section 7.2 to determine the form of overtopping for shore-normal (zero obliquity). Based upon the outcome of this, guidance under "non-impulsive conditions" or "impulsive conditions" should be followed.

For non-impulsive conditions, an adjusted version of Equation 7.3 should be used (Equation 7.15):

$$\frac{q}{\sqrt{gH_{m0}^3}} = 0.04 \exp\left(-\frac{2.6}{\gamma}\frac{R_c}{H_{m0}}\right) \qquad 7.15$$

where γ is the reduction factor for angle of attack and is given by

$$\begin{aligned}\gamma &= 1 - 0.0062\beta &&\text{for } 0° < \beta < 45° \\ \gamma &= 0.72 &&\text{for } \beta \geq 45°\end{aligned} \qquad 7.16$$

and β is the angle of attack relative to the normal, in degrees.

For conditions that would be identified as impulsive for normal ($\beta = 0°$) wave attack, a more complex picture emerges (NAPP et al., 2004). Diminished incidence of impulsive overtopping is observed with increasing obliquity (angle β) of wave attack. This results not only in reductions in mean discharge with increasing β but also, for $\beta \geq 60°$, a switch back over to the functional form observed for non-impulsive conditions (i.e. a move away from a power-law decay such as Equation 7.6 to an exponential one such as Equation 7.3).

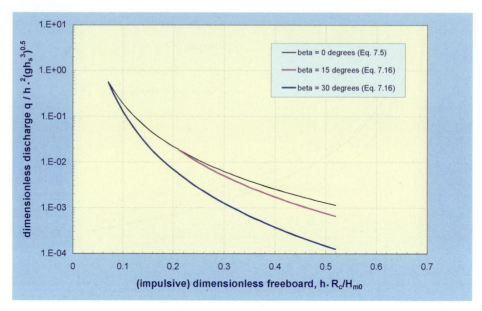

Fig. 7.16: Overtopping of vertical walls under oblique wave attack

For **probabilistic design or comparison with measurements,** the mean predictions should be used (Equation 7.17) should be used. Data only exist for the discrete values of obliquity listed.

for β = 15° ; $h_* \dfrac{R_c}{H_{m0}} \geq 0.2$ $\qquad \dfrac{q}{h_*^2 \sqrt{gh_s^3}} = 5.8 \times 10^{-5} \left(h_* \dfrac{R_c}{H_{m0}} \right)^{-3.7}$

for β = 15° ; $h_* \dfrac{R_c}{H_{m0}} < 0.2$ $\qquad$ as per impulsive β = 0° (Eq. 7.6).

for β = 30° ; $h_* \dfrac{R_c}{H_{m0}} \geq 0.07$ $\qquad \dfrac{q}{h_*^2 \sqrt{gh_s^3}} = 8.0 \times 10^{-6} \left(h_* \dfrac{R_c}{H_{m0}} \right)^{-4.2}$ 7.17

for β = 60° ; $h_* \dfrac{R_c}{H_{m0}} \geq 0.07$ $\qquad$ as per non-impulsive β=60° (Eq. 7.16).

Significant spatial variability of overtopping volumes along the seawall under oblique wave attack are observed/measured in physical model studies. For **deterministic design,** Equation 7.18 should be used, as these give estimates of the "worst case" conditions at locations along the wall where the discharge is greatest.

for β = 15° ; $h_* \dfrac{R_c}{H_{m0}} \geq 0.2$ $\qquad$ as per impulsive β = 0° (Eq.7.7)

for β = 30° ; $h_* \dfrac{R_c}{H_{m0}} \geq 0.07$ $\qquad$ as per impulsive β = 15° (Eq.7.17) 7.18

for β = 60° ; $h_* \dfrac{R_c}{H_{m0}} \geq 0.07$ $\qquad$ as per non-impulsive β = 0° (Eq.7.4)

7.3.5 Effect of bullnose and recurve walls

Designers of vertical seawalls and breakwaters have often included some form of seaward overhang (*recurve/parapet/wave return wall/bullnose*) as part of the structure with the design motivation of reducing wave overtopping by deflecting back seaward uprushing water (eg Fig. 7.18). The mechanisms determining the effectiveness of a recurve are complex and not yet fully described. The guidance presented here is based upon physical model studies (KORTENHAUS et al., 2003; PEARSON et al., 2004).

Parameters for the assessment of overtopping at structures with bullnose/recurve walls are shown in Fig. 7.19.

Fig. 7.17: An example of a modern, large vertical breakwater with wave return wall (left) and cross-section of an older seawall with recurve (right)

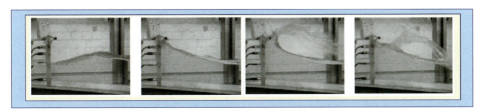

Fig. 7.18: A sequence showing the function of a parapet/wave return wall in reducing overtopping by redirecting the uprushing water seaward (back to right)

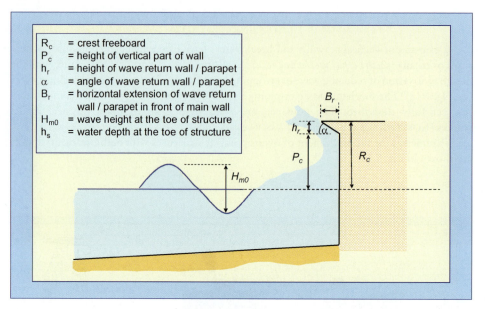

Fig. 7.19: Parameter definitions for assessment of overtopping at structures with parapet/wave return wall

Two conditions are distinguished;
- the familiar case of the parapet/bullnose/recurve overhanging seaward ($\alpha < 90°$), and
- the case where a wall is chamfered backwards at the crest (normally admitting *greater* overtopping ($\alpha > 90°$).

For the latter, chamfered wall case, Cornett influence factors γ should be applied to Franco's equation for non-impulsive mean discharge (Equation 7.19) with a value of γ selected as shown (CORNETT et al., 1999).

$$\frac{q}{\left(gH_{m0}^3\right)^{0.5}} = 0.2\exp\left(-\frac{4.3}{\gamma}\frac{R_c}{H_{m0}}\right)$$

7.19

$\gamma = 1.01$ for $\alpha = 120°$
$\gamma = 1.13$ for $\alpha = 135°$
$\gamma = 1.07$ for $\alpha = 150°$

For the familiar case of overhanging parapet/recurve/bullnose, the effectiveness of the recurve/parapet in reducing overtopping is quantified by a factor k defined as

$$k = \frac{q_{\text{with\_recurve}}}{q_{\text{without\_recurve}}}$$

7.20

The decision chart in Fig. 7.20 can then be used to arrive at a value of k, which in turn can be applied by multiplication to the mean discharge predicted by the most appropriate method for the plain vertical wall (with the same R_c, h_s etc.). The decision chart shows three levels of decision;

- whether the parapet is angled seaward or landward;
- if seaward ($\alpha < 90°$), whether conditions are in the small (left box), intermediate (middle box) or large (right box) reduction regimes;
- if in the regime of largest reductions (greatest parapet effectiveness; $R_c/H_{m0} \geq R_0^* + m^*$), which of three further sub-regimes (for different R_c/h_s) is appropriate.

Given the level of scatter in the original data and the observation that the methodology is not securely founded on the detailed physical mechanisms/processes, it is suggested that it is impractical to design for $k < 0.05$, i.e. reductions in mean discharges by factors of greater than 20 cannot be predicted with confidence. If such large (or larger) reductions are required, a detailed physical model study should be considered.

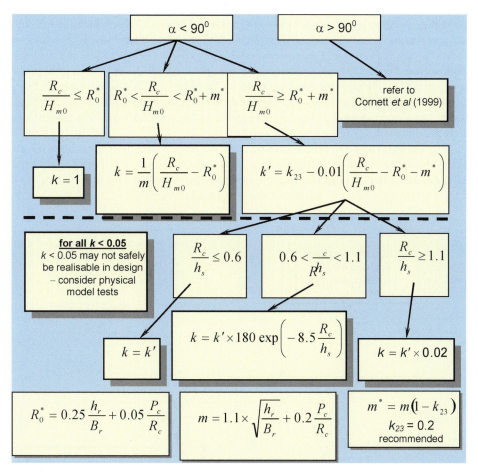

Fig. 7.20: "Decision chart" summarising methodology for tentative guidance. Note that symbols R_0^*, k_{23}, m and m^* used (only) at intermediate stages of the procedure are defined in the lowest boxes in the figure. Please refer to text for further explanation

7.3.6 Effect of wind

Wind may affect overtopping processes and thus discharges by:
- changing the shape of the incident wave crest at the structure resulting in a possible modification of the dominant regime of wave interaction with the wall;
- blowing up-rushing water over the crest of the structure (for an onshore wind, with the reverse effect for an offshore wind) resulting in possible modification of mean overtopping discharge and wave-by-wave overtopping volumes;
- modifying the physical form of the overtopping volume or jet, especially in terms of its aeration and break-up resulting in possible modification to post-overtopping characteristics such as throw speed, landward distribution of discharge and any resulting post-overtopping loadings (e.g. downfall pressures).

The modelling of any of these effects in small-scale laboratory tests presents very great difficulties owing to fundamental barriers to the simultaneous scaling of the wave-structure and water-air interaction processes. Very little information is available to offer guidance on effect (1) – the reshaping of the incident waves. Comparisons of laboratory and field data (both with and without wind) have enabled some upper (conservative) bounds to be placed upon effect (2) – the intuitive wind-assistance in "pushing" of up-rushing water landward across the crest. These are discussed immediately below. Discussion of effect (3) – modification to "post-overtopping" processes – is reserved for Sections 7.5.3 and 7.5.4 (on distributions and downfalling pressures respectively).

For vertical structures, several investigations on vertical structures have suggested different adjustment factors f_{wind} ranging from 30 % to 40 % to up to 300 % (Fig. 7.21) either using a paddle wheel or large fans to transport uprushing water over the wall.

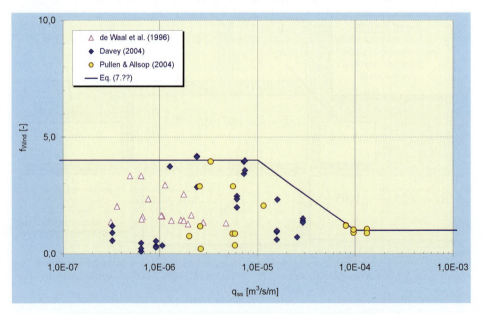

Fig. 7.21: Wind adjustment factor f_{wind} plotted over mean overtopping rates q_{ss}

When these tests were revisited a simple adjustment factor was proposed for the mean discharge based upon small-scale tests q_{ss}, which is already scaled up by appropriate scaling to full-scale (see also de ROUCK et al., 2005).

$$f_{wind} = \begin{cases} 4.0 & \text{for } q_{ss} \leq 10^{-5} \text{ m}^3/\text{s/m} \\ 1.0 + 3.(-\log q_{ss} - 4) & \text{for } 10^{-5} < q_{ss} < 10^{-4} \text{ m}^3/\text{s/m} \\ 1.0 & \text{for } q_{ss} \geq 10^{-4} \text{ m}^3/\text{s/m} \end{cases} \qquad 7.21$$

From Equation 7.21 it becomes clear that the influence of wind only gets important for very low overtopping rates below q_{ss} = 0.1 l/s/m. Hence, in many practical cases, the influence of wind may be disregarded. The mean overtopping discharge including wind becomes

$$q_{\text{with wind}} = f_{wind} \times q_{ss} \qquad 7.22$$

7.3.7 Scale and model effect corrections

Tests in a large-scale wave channel (Fig. 7.22) and field measurements (Fig. 7.23) have demonstrated that with the exception of wind effect (Section 7.3.6), results of overtopping measurements in small-scale laboratory studies may be securely scaled to full-scale under non-impulsive and impulsive overtopping conditions (PEARSON et al., 2002; PULLEN et al., 2004).

No information is yet available on the scaling of small-scale data under conditions where broken wave attack dominates. Comparison of measurements of wave loadings on vertical structures under broken wave attack at small-scale and in the field suggests that prototype loadings will be *over-estimated* by small-scale tests in the presence of highly-aerated broken waves. Thus, although the methods presented for the assessment of overtopping discharges under broken wave conditions given in Section 7.3.1 have not been verified at large-scale or in the field, any scale correction is expected to give a *reduction* in predicted discharge.

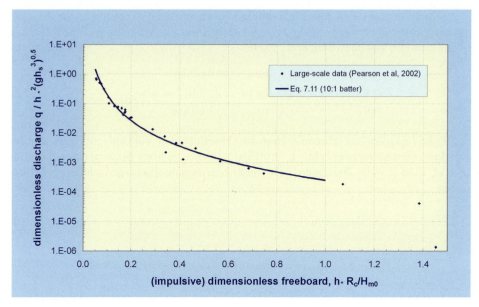

Fig. 7.22: Large-scale laboratory measurements of mean discharge at 10:1 battered wall under impulsive conditions showing agreement with prediction line based upon small-scale tests (Equation 7.12)

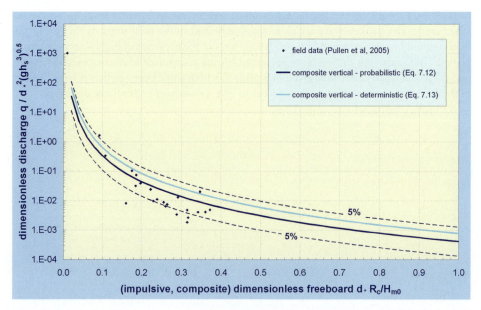

Fig. 7.23: Results from field measurements of mean discharge at Samphire Hoe, UK, plotted together with Equation 7.13

7.4 Overtopping volumes

7.4.1 Introduction

While the prediction of mean discharge (Section 7.3) offers the information required to assess whether overtopping is slight, moderate or severe, and make a link to any possible flooding that might result, the prediction of the volumes associated with individual wave events can offer an alternative (and often more appropriate) measure for the assessment of tolerable overtopping levels and possible direct hazard. First, a method is given for the prediction of maximum overtopping volumes expected associated with individual wave events for plain vertical structures under perpendicular wave attack (Section 7.4.2). This method is then extended to composite (bermed) structures (Section 7.4.3) and to conditions of oblique wave attack (Section 7.4.4). Finally, a short section on scale effects is included (Section 7.4.5). Also refer to Section 4.2.2.

The methods given for perpendicular wave attack are the same as those given previously in UK guidance (EA/BESLEY, 1999). Only the extension to oblique wave attack is new.

7.4.2 Overtopping volumes at plain vertical walls

The first step in the estimation of a maximum expected individual wave overtopping volume is to estimate the number of waves overtopping (N_{ow}) in a sequence of N_w incident waves.

For **non-impulsive conditions**, this was found to be well-described by (FRANCO et al., 1994)

$$N_{ow} = N_w \exp\left\{-1.21\left(\frac{R_c}{H_{m0}}\right)^2\right\} \quad \text{(for } h_* > 0.3\text{)} \qquad 7.23$$

(arising from earlier tests on sloping structures in which situation the number of overtopping waves was directly linked to run-up, in turn linked to a Rayleigh-distributed set of incident wave heights).

Under **impulsive conditions**, N_{ow} is better described by (EA/BESLEY, 1999)

$$N_{ow} = 0.031 N_w \times \frac{H_{m0}}{h_* R_c} \quad \text{(for } h_* > 0.3\text{)} \qquad 7.24$$

where $h_* R_c / H_{m0}$ is the dimensionless freeboard parameter for impulsive conditions (Equation 7.1).

The distribution of individual overtopping volumes in a sequence is generally well-described by a two-parameter Weibull distribution (also refer to Section 4.2.2);

$$P_V = 1 - \exp\left\{-\left(\frac{V}{a}\right)^b\right\} \qquad 7.25$$

where P_V is the probability that an individual event volume will not exceed V. a and b are Weibull "shape" and "scale" parameters respectively. Thus, to estimate the largest event in a wave sequence predicted to include (*e.g.*) N_{ow} = 200 overtopping events, V_{max} would be found by taking P_V = 1/200 = 0.005. Equation 7.25 can then be rearranged to give

$$V_{max} = a \, (\ln N_{ow})^{1/b} \qquad 7.26$$

The Weibull shape parameter a depends upon the average volume per overtopping wave V_{bar} where

$$V_{bar} = \frac{q T_{m-1,0} N_w}{N_{ow}} \qquad 7.27$$

For *non-impulsive conditions*, there is a weak steepness-dependency for the scale and shape parameters *a* and *b* (FRANCO (1996));

$$a = \begin{cases} 0.74 V_{bar} \\ 0.90 V_{bar} \end{cases} \quad b = \begin{cases} 0.66 & \text{for } s_{m-1,0} = 0.02 \\ 0.82 & \text{for } s_{m-1,0} = 0.04 \end{cases} \quad (\text{for } h_* > 0.3) \qquad 7.28$$

For *impulsive conditions*, (EA/BESLEY, 1999; PEARSON et al., 2002);

$$a = 0.92 V_{bar} \quad b = 0.85 \qquad (\text{for } h_* < 0.3) \qquad 7.29$$

The effectiveness of the predictor under impulsive conditions can be gauged from Fig. 7.24.

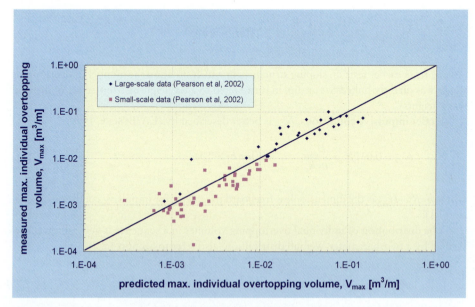

Fig. 7.24: Predicted and measured maximum individual overtopping volume – small- and large-scale tests (PEARSON et al., 2002)

7.4.3 Overtopping volumes at composite (bermed) structures

There is very little information available specifically addressing wave-by-wave overtopping volumes at composite structures. The guidance offered by EA/BESLEY (1999) remains the best available. No new formulae or Weibull a, b values are known so, for the purposes of maximum overtopping volume prediction, the methods for plain vertical walls (Section 7.4.2) are used. The key discriminator is that composite structures whose mound is sufficiently small to play little role in the overtopping process are treated as *plain vertical, non-impulsive*, whereas those with large mounds are treated as *plain vertical, impulsive*.

For this purpose, the significance of the mound is assessed using the "impulsiveness" parameter for composite structures, $d*$ (Equation 7.2). "Small mound" is defined as $d* > 0.3$, with $d* < 0.3$ being "large mound".

7.4.4 Overtopping volumes at plain vertical walls under oblique wave attack

For *non-impulsive* conditions, an adjusted form of Equation 7.23 is suggested (FRANCO et al., 1994);

$$N_{ow} = N_w \exp\left\{-\frac{1}{C^2}\left(\frac{R_c}{H_{m0}}\right)^2\right\} \quad \text{(for } h_* > 0.3\text{)} \qquad 7.30$$

C is given by

$$C = \begin{cases} 0.91 & \text{for } \beta = 0° \\ 0.91 - 0.00425\beta & \text{for } 0° < \beta < 40° \\ 0.74 & \text{for } \beta \geq 40° \end{cases} \quad \text{(for } h_* > 0.3\text{)} \qquad 7.31$$

For *impulsive conditions* (as determined for perpendicular i.e. $\beta = 0°$ wave attack), the procedure is the same as for perpendicular ($\beta = 0°$) wave attack, but different formulae should be used for estimating the number of overtopping waves (N_{ow}) and Weibull shape and scale parameters – see Table 7.2 (NAPP et al., 2004).

Table 7.2: Summary of prediction formulae for individual overtopping volumes under oblique wave attack. Oblique cases valid for $0.2 < h* R_c / H_{m0} < 0.65$. For $0.07 < h* R_c / H_{m0} < 0.2$, the $\beta = 0°$ formulae should be used for all β

$\beta = 15°$	$\beta = 30°$	$\beta = 60°$
$N_{ow} = 0.01 N_w \times \left(\frac{H_{m0}}{h_* R_c}\right)^{-1.6}$	$N_{ow} = 0.01 N_w \times \left(\frac{H_{m0}}{h_* R_c}\right)^{-1.4}$	treat as non-impulsive
$a = 1.06 V_{bar}$	$a = 1.04 V_{bar}$	treat as non-impulsive
$b = 1.18$	$b = 1.27$	treat as non-impulsive

7.4.5 Scale effects for individual overtopping volumes

Measurements from large-scale laboratory tests indicate that formulae for overtopping volumes, based largely upon small-scale physical model studies, scale well (Fig. 7.24) (PEARSON et al., 2002). No data from the field is available to support "scale-ability" from large-scale laboratory scales to prototype conditions.

7.5 Overtopping velocities, distributions and down-fall pressures

7.5.1 Introduction to post-overtopping processes

There are many design issues for which knowledge of just the mean and/or wave-by-wave overtopping discharges/volumes are not sufficient, e.g.
- assessment of direct hazard to people, vehicles and buildings in the zone immediately landward of the seawall;
- assessment of potential for damage to elements of the structure itself (e.g. crown wall; crown deck; secondary defences);

The appreciation of the importance of being able to predict more than overtopping discharges and volumes has led to significant advances in the description and quantification of what can be termed "post-overtopping" processes. Specifically, the current state of prediction tools for
- the speed of an overtopping jet (or "throw velocity");
- the spatial extent reached by (impulsive) overtopping volumes, and
- the pressures that may arise due to the downfalling overtopped jet impacting on the structure's crown deck.

7.5.2 Overtopping throw speeds

Studies at small-scale based upon video footage (Fig. 7.25) suggest that the vertical speed with which the overtopping jet leaves the crest of the structure (u_z) may be estimated as

$$u_z \approx \begin{cases} 2 \text{ to } 2.5 \times c_i & \text{for non-impulsive conditions} \\ 5 \text{ to } 7 \times c_i & \text{for impulsive conditions} \end{cases} \quad 7.32$$

where $c_i = \sqrt{gh_s}$ is the inshore wave celerity (BRUCE et al., 2002).

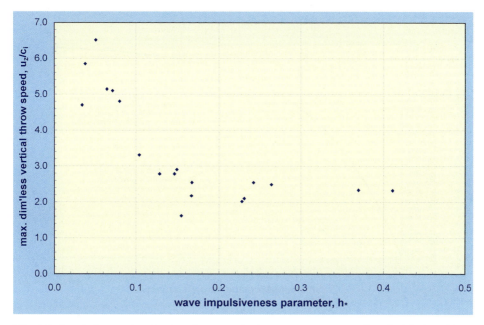

Fig. 7.25: Speed of upward projection of overtopping jet past structure crest plotted with "impulsiveness parameter" h_* (after BRUCE et al., 2002)

7.5.3 Spatial extent of overtopped discharge

The spatial distribution of overtopped discharge may be of interest in determining zones affected by direct wave overtopping hazard (to people, vehicles, buildings close behind the structure crest, or to elements of the structure itself).

Under green water (non-impulsive) conditions, the distribution of overtopped water will depend principally on the form of the area immediately landward of the structures crest (slopes, drainage, obstructions etc.) and no generic guidance can be offered (though see Section 7.5.2 for information of speeds of overtopping jets).

Under violent (impulsive) overtopping conditions, the idea of spatial extent and distribution has a greater physical meaning – where does the airborne overtopping jet come back to the level of the pavement behind the crest? The answer to this question however will (in general) depend strongly upon the local wind conditions. Despite the difficulty of directly linking a laboratory wind speed to its prototype equivalent (see Section 7.3.6) laboratory tests have been used to place an upper bound on the possible wind-driven spatial distribution of the "fall back to ground" footprint of the violently overtopped volumes (PULLEN et al., 2004 and BRUCE et al., 2005). Tests used large fans to blow air at gale-force speeds (up to 28 ms^{-1}) *in the laboratory*. The resulting landward distributions for various laboratory wind speeds are shown in Fig. 7.26. The lower (conservative) envelope of the data give the approximate guidance that 95 % of the violently-overtopped discharge will land within a distance of $0.25 \times L_o$, where L_o is the offshore (deep water) wavelength.

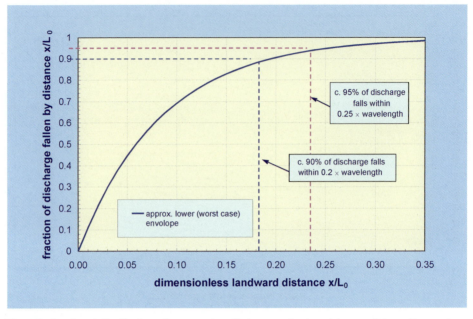

Fig. 7.26: Landward distribution of overtopping discharge under impulsive conditions. Curves show proportion of total overtopping discharge which has landed within a particular distance shoreward of seaward crest

7.5.4 Pressures resulting from downfalling water mass

Wave impact pressures on the crown deck of a breakwater have been measured in small- and large-scale tests (BRUCE et al., 2001; WOLTERS et al., 2005). These impacts are the result of an impacting wave at the front wall of the breakwater generating an upwards jet which in turn falls back onto the crown deck of the structure. Small-scale tests suggest that local impact pressure maxima on the crown deck are smaller than *but of the same order of magnitude* as wave impact pressures on the front face. For high-crested structures ($R_c / H_{m0} > 0.5$), pressure maxima were observed to occur within a distance of ~ $1.5 \times H_{m0}$ behind the seaward crest. For lower-crested structures ($R_c / H_{m0} < 0.5$) this distance was observed to increase to ~ $2 \times H_{m0}$. Over all small-scale tests, pressure maxima were measured over the range

$$2 < \frac{p_{1/250}}{\rho g H_{m0}} < 17 \quad \text{with a mean value of 8} \qquad 7.33$$

The largest downfall impact pressure measured in large-scale tests was 220 kPa (with a duration of 0.5 ms). The largest downfall pressures were observed to result from overtopping jets thrown upwards by very-nearly breaking waves (the "flip through" condition). Although it might be expected that scaling small-scale impact pressure data would over-estimate pressure maxima at large scale, approximate comparisons between small- and large-scale test data suggest that the agreement is good.

7.6 Uncertainties

Wave overtopping formulae for vertical and steep seawalls depend on the type of wall which is overtopped and the type of wave breaking at the wall. The wave overtopping formulae used are however similar to the ones used for sloping structures such as dikes and rubble mound structures. Therefore, again the same procedure is suggested as used already for Sections 5.7 and 6.3.7.

The uncertainty in crest height variation for vertical structures is different from sloping structures and should be set to about 0.04 m. All uncertainties related to waves and water levels will remain as discussed within Section 5.7. Similarly, the results of these additional uncertainties have little influence on the results using the model uncertainty only. This is evident from (e.g.) Fig. 7.10 for impulsive conditions at a plain vertical wall.

Resulting probabilistic and deterministic design parameters are summarised in Table 7.3.

Table 7.3: Probabilistic and deterministic design parameters for vertical and battered walls

Type of wall	Type of breaking	Type of formula	Probabilistic par.	Deterministic par.
Plain vertical	non-impulsive	Eq. 7.4	$a = 0.04$; $b = -2.62$	$a = 0.04$; $b = -1.80$
	impulsive	Eq. 7.6	$a = 1.48 \cdot 10^{-4}$; $b = -3.09$	$a = 2.77 \cdot 10^{-4}$; $b = -3.09$
	emergent toe, impulsive	Eq. 7.10	$a = 2.72 \cdot 10^{-4}$; $b = -2.69$	$a = 3.92 \cdot 10^{-4}$; $b = -2.69$
Composite	non-impulsive	Eq. 7.4	$a = 0.016$; $b = -3.28$	$a = 0.016$; $b = -2.75$
	impulsive	Eq. 7.12	$a = 4.10 \cdot 10^{-4}$; $b = -2.91$	$a = 7.18 \cdot 10^{-4}$; $b = -2.91$

It is noteworthy that only uncertainties for mean wave overtopping rates have been considered here (as per previous sections dealing with uncertainties). Other methods discussed in this chapter have not been considered per se, but can be dealt with using the principal procedure as discussed in Section 1.5.4.

Glossary

Armour	Protective layer of rock or concrete units
Composite sloped seawall	A sloped seawall whose gradient changes
Composite vertical wall	A structure made up of two component parts, usually a caisson type structure constructed on a rubble mound foundation
Crown wall	A concrete super-structure located at the crest of a sloping seawall
Deep water	Water so deep that that waves are little affected by the seabed. Generally, water deeper than one half the surface wavelength is considered to be deep
Depth limited waves	Breaking waves whose height is limited by the water depth
Crest Freeboard	The height of the crest above still water level
Impulsive waves	Waves that tend to break onto the seawall
Mean overtopping discharge	The average flow rate passing over the seawall
Mean wave period	The average of the wave periods in a random sea state
Model effects	Model effects occur due to the inappropriate set-up of the model and the incorrect reproduction of the governing forces, the boundary conditions, the measurement system and the data analysis
Normal wave attack	Waves that strike the structure normally to its face
Oblique wave attack	Waves that strike the structure at an angle
Overflow discharge	The amount of water passing over a structure when the water level in front of the structure is higher than the crest level of the structure
Peak overtopping discharge	The largest volume of water passing over the structure in a single wave
Reflecting waves	Waves that hit the structure and are reflected seaward with little or no breaking
Return period	The average length of time between sea states of a given severity
Run-up	The rush of water up a structure or beach as a result of wave action
Scale effects	Scale effects occur due to the inability to scale all relevant forces from prototype to model scale
Sea dike	Earth structure with a sand core covered by clay, sometimes covered by asphalt or concrete

Shallow Water	Water of such a depth that surface waves are noticeably affected by bottom topography. Customarily water of depth less than half the surface wavelength is considered to be shallow
Significant wave height	The average height of the highest of one third of the waves in a given sea state
Toe	The relatively small mound usually constructed of rock armour to support or key-in armour layer
Tolerable overtopping discharge	The amount of water passing over a structure that is considered safe
Wave return wall	A wall located at the crest of a seawall, which is designed to throw back the waves
Wave steepness	The ratio of the height of the waves to the wave length

Notation

A_c	= armour crest freeboard of structure	[m]
B	= berm width, measured horizontally	[m]
B_t	= width of toe of structure	[m]
B_h	= width of horizontally schematised berm	[m]
B_{ov}	= longitudinal extension of overtopping front	[m]
B_r	= width (seaward extension) in front of main vertical wall of recurve/parapet/wave return wall section	[m]
c	= wave celerity at structure toe	[m/s]
C_r	= average reflection coefficient (= $\sqrt{m_{0,r}}/\sqrt{m_{0,i}}$)	[– or %]
CF	= Complexity-Factor of structure section, gives an indication of the complexity of the structure section, can adopt the values 1, 2, 3 or 4	[-]
D_{n50}	= nominal diameter of rock	[m]
D_n	= nominal diameter of concrete armour unit	[m]
$D(f,\theta)$	= directional spreading function, defined as:	[°]
	$S(f, \theta) = S(f) \cdot D(f,\theta)$ met $\int_0^{2\pi} D(f,\theta)d\theta = 0$	
f	= frequency	[Hz]
f_p	= spectral peak frequency	
	= frequency at which $S\eta(f)$ is a maximum	[Hz]
f_b	= width of a roughness element (perpendicular to structure axis)	[m]
f_h	= height of a roughness element	[m]
f_L	= centre-to-centre distance between roughness elements	[m]
g	= acceleration due to gravity (= 9,81)	[m/s²]
G_c	= width of structure crest	[m]
h	= water depth at toe of structure	[m]
h_b	= water depth on berm (negative means berm is above S.W.L.)	[m]
h_{deep}	= water depth in deep water	[m]
h_r	= height of recurve/parapet/wave return wall section at top of vertical wall	[m]
h_t	= water depth on toe of structure	[m]
H	= wave height	[m]
$H_{1/x}$	= average of highest 1/x th of wave heights	[m]
$H_{x\%}$	= wave height exceeded by x% of all wave heights	[m]
H_s	= significant wave height defined as highest one-third of wave heights	
	= $H_{1/3}$	[m]
H_{m0}	= estimate of significant wave height from spectral analysis	
	= $4\sqrt{m_0}$	[m]
$H_{m0\ deep}$	= H_{m0} determined at deep water	[m]
$H_{m0\ toe}$	= H_{m0} determined at toe of structure	[m]
k	= angular wave number (= $2\pi/L$)	[rad/m]
k	= multiplier for mean discharge giving effect of recurve wall (Chapter 7)	[-]
k', k_{23}	= dimensionless parameters used (only) in intermediate stage of calculation of reduction factor for recurve walls (Chapter 7)	[-]

L_{berm}	= horizontal length between two points on slope, $1.0\ H_{m0}$ above and $1.0\ H_{m0}$ below middle of the berm	[m]
L_{slope}	= horizontal length between two points on slope, $Ru_{2\%}$ above and $1.5\ H_{m0}$ below S.W.L.	[m]
L	= wave length measured in direction of wave propagation	[m]
L_{0p}	= peak wave length in deep water = $gT^2_p/2\pi$	[m]
L_{0m}	= mean wave length in deep water = $gT^2_m/2\pi$	[m]
L_0	= deep water wave length based on $T_{m-1,0} = gT^2_{m-1,0}/2\pi$	[m]
m	= slope of the foreshore: 1unit vertical corresponds to m units horizontal	[-]
m^*, m	= dimensionless parameters used (only) in intermediate stage of calculation of reduction factor for recurve walls (Chapter 7)	[-]
m_n	= $\int_{f_1}^{f_2} f^n S(f) df = n^{th}$ moment of spectral density	[m²/sⁿ]
	lower integration limit = f_1 = min(1/3.f_p, 0.05 full scale)	
	upper integration limit = $f_2 = 3.f_p$	
$m_{n,x}$	= n^{th} moment of x spectral density	[m²/sⁿ]
	x may be: i for incident spectrum	
	r for reflected spectrum	
N_{ow}	= number of overtopping waves	[-]
N_w	= number of incident waves	[-]
P(x)	= probability distribution function	
p(x)	= probability density function	
P_c	= height of vertical wall from SWL to bottom of recurve/parapet/wave return wall section (i.e. $P_c = R_c - h_r$)	[m]
P_V	= $P(\underline{V} \geq V)$ = probability of the overtopping volume $\underline{V}$ being larger or equal to V	[-]
P_{ow}	= probability of overtopping per wave = N_{ow}/N_w	[-]
q	= mean overtopping discharge per meter structure width	[m³/s/m]
R_c	= crest freeboard of structure	[m]
R_{cL}	= crest freeboard of structure landward side (relative to falling plane)	[m]
RF	= Reliability-Factor of test, gives an indication of the reliability of the test, can adopt the values 1, 2, 3 or 4	[-]
R_0^*	= dimensionless length parameter used (only) in intermediate stage of calculation of reduction factor for recurve walls (Chapter 7)	[-]
R_u	= run-up level, vertical measured with respect to the S.W.L.	[m]
$R_{u2\%}$	= run-up level exceeded by 2 % of incident waves	[m]
R_{us}	= run-up level exceeded by 13.6 % of incident waves	[m]
s	= wave steepness = H/L	[-]
s_{0p}	= wave steepness with L_0, based on $T_p = H_{m0}/L_{0p} = 2\pi H_{m0}/(gT^2_p)$	[-]
s_{0m}	= wave steepness with L_0, based on $T_m = H_{m0}/L_{0m} = 2\pi H_{m0}/(gT^2_m)$	[-]
s_0	= wave steepness with L_0, based on $T_{m-1,0} = H_{m0}/L_0 = 2\pi H_{m0}/(gT^2_{m-1,0})$	[-]
$S_{\eta,i}(f)$	= incident spectral density	[m²/Hz]
$S_{\eta,r}(f)$	= reflected spectral density	[m²/Hz]

$S(f, \theta)$	= directional spectral density	[(m²/Hz)/]
t	= variable of time	[s]
T	= wave period	[s]
$T_{H1/x}$	= average of the periods of the highest 1/x th of wave heights	[s]
T_m	= average wave period defined either as:	
	$\bar{T}$ = average wave period from time-domain analysis	[s]
	$T_{mi,j}$ = average wave period calculated from spectral moments, e.g.:	[s]
$T_{m0,1}$	= average wave period defined by m_0/m_1	[s]
$T_{m0,2}$	= average wave period defined by $\sqrt{m_0/m_2}$	[s]
$T_{m-1,0}$	= average wave period defined by m_{-1}/m_0	[s]
$T_{m-1,0\ deep}$	= $T_{m-1,0}$ determined at deep water	[s]
$T_{m-1,0\ toe}$	= $T_{m-1,0}$ determined at the toe of the structure	[s]
$T_{m\ deep}$	= T_m determined at deep water	[s]
$T_{m\ toe}$	= T_m determined at the toe of the structure	[s]
T_p	= spectral peak wave period = $1/f_p$	[s]
$T_{p\ deep}$	= T_p determined at deep water	[s]
$T_{p\ toe}$	= T_p determined at the toe of the structure	[s]
T_R	= record length or return period of event	[s]
T_s	= $T_{H1/3}$ = significant wave period	[s]
V	= volume of overtopping wave per unit crest width	[m³/m]
V_{max}	= maximum overtopping volume per wave per unit crest width	[m³/m]
v	= velocity of overtopping jet at wall detachment point	[m/s]
X	= landward distance of falling overtopping jet from rear edge of wall	[m]
X_{max}	= maximum landward distance of falling overtopping jet from rear edge of wall	[m]
X_{qmax}	= landward distance of max mean discharge	[m]
X_{Vmax}	= landward distance of max overtopping volume per wave	[m]
W_s	= wind speed ($W_s \times \cos W_d$ = wind speed onshore component normal to structure)	[m/s]
W_d	= wind direction-angle of wind attack relative to normal on structure	[°]
α	= angle between overall structure slope and horizontal	[°]
α	= angle of parapet / wave return wall above seaward horizontal	[°]
α_B	= angle that sloping berm makes with horizontal	[°]
α_u	= angle between structure slope upward berm and horizontal	[°]
α_d	= angle between structure slope downward berm and horizontal	[°]
α_{excl}	= mean slope of structure calculated without contribution of berm	[°]
α_{incl}	= mean slope of structure calculated with contribution of berm	[°]
α_{wall}	= angle that steep wall makes with horizontal	[°]
β	= angle of wave attack relative to normal on structure	[°]
$\eta(t)$	= surface elevation with respect to S.W.L.	[m]
γ_b	= correction factor for a berm	[-]
γ_f	= correction factor for the permeability and roughness of or on the slope	[-]
γ_β	= correction factor for oblique wave attack	[-]

γ_v	= correction factor for a vertical wall on the slope	[-]
ξ_o	= breaker parameter based on s_o (= $\tan\alpha/s_o^{1/2}$)	[-]
ξ_{om}	= breaker parameter based on s_{om}	[-]
ξ_{op}	= breaker parameter based on s_{op}	[-]
$\mu_{(x)}$	= mean of measured parameter x with normal distribution	[..]
$\sigma_{(x)}$	= standard deviation of measured parameter x with normal distribution	[..]
θ	= direction of wave propagation	[°]
ω	= angular frequency = $2\pi f$	[rad/s]

References

Abernethy, R. & Olliver, G. 2002 Effects of modelling long and short crested seas on overtopping of a long, vertical faced breakwater, Proc 28th Int. Conf. Coastal Engng (ASCE), Cardiff, pp 2299–2311, World Scientific, ISBN 981-238-238-0.

Alberti, P., Bruce, T. & Franco, L. 2001 Wave transmission behind vertical walls due to overtopping. Paper 21 in Proc. Conf. Shorelines, Structures & Breakwaters, September 2001, pp 269–282, ICE, London.

Allsop, N. W. H., Bruce, T., Pearson, J., Alderson, J. S. & Pullen, T. 2003 Violent wave overtopping at the coast, when are we safe? Proc. Conf. on Coastal Management 2003, pp 54–69, ISBN 0-7277-3255-2, publn. Thomas Telford, London.

Allsop, N. W. H., Bruce, T., Pearson, J., Franco, L., Burgon, J. & Ecob, C. 2004 Safety under wave overtopping – how overtopping processes and hazards are viewed by the public. Proc. 29th Int. Conf. on Coastal Engng. Lisbon. pp 4263–4274.

Allsop, N. W. H., Franco, L., Bellotti, G., Bruce, T. & Geeraerts, J. 2005 Hazards to people and property from wave overtopping at coastal structures. Proc. Conf. Coastlines, Structures and Breakwaters, 20–22 April 2005, ICE, London.

Allsop, N. W. H., Lihrmann, H., & Netherstreet, I. 2002 Wave breaking on/over steep slopes. Paper 16a in "Breakwaters, coastal structures & coastlines" ICE, ISBN 0-7277-3042-8, pp 215–218, publn Thomas Telford, London.

Allsop, N. W. H., Besley, P. & Madurini, L. 1995 Overtopping performance of vertical and composite breakwaters, seawalls and low reflection alternatives. Paper 4.7 in MCS Project Final Report, University of Hannover.

Allsop, N. W. H., Bruce, T., Pearson, J. & Besley, P. 2005 Wave overtopping at vertical and steep seawalls. Proc. ICE, Maritime Engineering, 158, 3, pp 103–114, ISSN 1741 7597

Aminti, P. L. & Franco, L. 1988 Wave overtopping on rubble mound breakwaters. Proc. 21st Int. Conf. on Coastal Engng. Torremolinos. 1988

Asbeck, W. F. Baron van, Ferguson, H. A., & Schoemaker, H. J. 1953 New designs of breakwaters and seawalls with special reference to slope protection. Proc. 18th Int. Nav. Congress, Rome, Sect. 2, Qu. 1, p. 174.

Banyard, L. & Herbert, D. M. 1995 The effect of wave angle on the overtopping of seawalls. HR Wallingford, Report SR396.

Battjes, J. A. & Groenendijk, H. W. 2000 Wave height distributions on shallow foreshores. Coastal Engineering Vol 40 pp 161–182, Elsevier Science, Rotterdam.

Battjes, J. A. & Stive, M. J. F. 1984 Calibration and verification of dissipation model for random breaking waves. Proc. 19th Int. Conf. on Coastal Engng. Houston.

Battjes, J. A. 1974 Surf Similarity. Proc. 14th Int. Conf. on Coastal Engng. Copenhagen. pp 466–480.

Battjes, J. A. 1971 Run-up distributions of waves breaking on slopes. Journal of the Waterways, Harbors and Coastal Engineering Division. Vol. 97. No. WW1. pp 91–114.

Berkeley-Thorn, R. & Roberts, A. C. 1981 Sea defence and coast protection works. Thomas Telford Ltd.

Besley, P. 1999 Overtopping of seawalls – design and assessment manual. R & D Technical Report W 178, Environment Agency, Bristol, ISBN 1-85705-069-X.

Besley, P., Stewart, T. & Allsop, N. W. H. 1998 Overtopping of vertical structures: new prediction methods to account for shallow water conditions. Proc. Conf. Coastlines, Structures and Breakwaters, ICE, London, Thomas Telford, ISBN 0-7277-3455-5

Bradbury, A. P. & Allsop, N. W. H. 1988 Hydraulic performance of breakwater crown walls. HR Wallingford, Report SR 146.

Brampton A. (Editor) 2002 Coastal defence – ICE design and practice guide. ISBN 0-7277-3005-3, Thomas Telford, London

Bruce, T., van der Meer, J. W., Franco L. & Pearson, J. 2006 A comparison of overtopping performance of different rubble mound breakwater armour. Proc. 30th Int. Conf. on Coastal Engng. San Diego.

Bruce, T., Allsop, N. W. H. & Pearson, J. 2001 Violent overtopping of seawalls – extended prediction methods. Paper 19 in Proc. Conf. Shorelines, Structures & Breakwaters, pp 245–255, ICE, London

Bruce, T., Pullen, T., Allsop, W. & Pearson, J. 2005 How far back from a seawall is safe? Spatial distributions of wave overtopping. Proc. International Conference on Coastlines, Structures and Breakwaters 2005, pp 166–176, ICE London, Thomas Telford, ISBN 0-7277-3455-5.

Bruce, T., Allsop, N. W. H., & Pearson, J. 2002 Hazards at coast and harbour seawalls – velocities and trajectories of violent overtopping jets. Proc. 28th Int. Conf. on Coastal Engng. Cardiff. pp 2216–2226.

Bruce, T., Franco, L., Alberti, P., Pearson, J. & Allsop, N. W. H. 2001 Violent wave overtopping: discharge throw velocities, trajectories and resulting crown deck loading. Proc. Ocean Wave Measurement and Analysis ('Waves 2001'), 2, pp 1783–1796, ASCE, New York, ISBN 0-7844-0604-9.

Bruce, T., Pearson, J. & Allsop, N. W. H. 2003 Violent wave overtopping – extension of prediction method to broken waves. Proc ,'Coastal Structures 2003", pp 619–630, ASCE, Reston, Virginia, ISBN 0-7844-0733-9.

Bruce, T., Pullen, T., Allsop, N. W. H. & Pearson, J. 2005 How far back from a seawall is safe? Spatial distributions of wave overtopping. Proc. Coastlines, Structures & Breakwaters 2005, pp 166–176, ICE London, Thomas Telford, ISBN 0-7277-3455-5.

Burcharth, H. F., Hawkins, S. J., Zanuttigh, B. & Lamberti, A 2007 Environmental Design Guidelines for Low Crested Coastal Structures. Elsevier, ISBN-13:978-0-08-044951-7.

Calabrese, M. 1998 Onset of breaking in front of vertical and composite breakwaters. Proc. 8th Int. Conf. ISOPE. Montreal, pp 590 – 595, ISBN 1-880653-34-6, publn ISOPE, San Francisco.

Calabrese, M. 1999 Occurrence of breaking on vertical breakwaters. Proc. 9th Int. Conf. ISOPE. Brest, pp 429–433, ISBN 1-880653-43-5, publn ISOPE, San Francisco.

CEM/Burcharth, H, F. & Hughes, S. A. 2002 Fundamentals of Design. In: Vincent, L., and Demirbilek, Z. (editors), Coastal Engineering Manual, Part VI, Design of Coastal Project Elements. Chapter VI-5-2, Engineer Manual 1110-2-1100, U.S. Army Corps of Engineers, Washington, DC.

CIRIA/CUR. 1991 Manual on the use of rock in coastal and shoreline engineering. CIRIA special publication 83, CUR Report 154, Simm J. D.(Editor), 1991.

Clarke, S., Dodd, N. & Damgaard, J. 2003 Flow on and in a porous beach. ASCE J. Waterway, Port, Coast & Ocean Eng., in review.

CLASH. Crest Level Assessment of coastal Structures by full scale monitoring, neural network prediction and Hazard analysis on permissible wave overtopping. Fifth Framework Programme of the EU, Contract n. EVK3-CT-2001-00058. www.clash-eu.org.

Cornett, A. & Mansard, E. 1994 Wave Stresses on rubble mound armour. Proc. 24th Int. Conf. on Coastal Engng. Kobe pp 986–1000.

Cornett, A., Li, Y. & Budvietas, A. 1999 Wave overtopping at chamfered and overhanging vertical structures. Proc. International Workshop on Natural Disasters by Storm Waves and Their Reproduction in Experimental Basins, Kyoto, Japan.

Cornett, A., Li, Y. & Budvietas, A. 1999 Wave overtopping at chamfered and overhanging vertical structures. Proc. International Workshop on Natural Disasters by Storm Waves and Their Reproduction in Experimental Basins, Kyoto, Japan.

DELOS. Environmental Design of Low Crested Coastal Defence Structures. Fifth Framework Programme of the EU, Contract n. EVK3-CT-00041. www.delos.unibo.it.

Daemrich, K.F.; Meyering, J.; Tack, G.; Zimmermann, C. 2006 Overtopping at vertical walls and parapeters - regular wave tests for irregular simulation. Proc. of the First Int. Conf. on the Application of physical modelling to Port and Coastal Protection. CoastLab. Porto. Portugal.

Daemrich, K.F.Meyering, J.; Ohle, N.; Zimmermann, C. (2006) Irregular wave overtopping at vertical walls – learning from regular wave tests. Proc. 30th Int. Conf. on Coastal Engng. San Diego. California.

De Rouck , J., Geeraerts, J., Troch, P., Kortenhaus, A., Pullen, T. & Franco, L. 2005 New results on scale effects for wave overtopping at coastal structures. Proc. Coastlines, Structures & Breakwaters 2005, pp 29–43, ICE London, Thomas Telford, ISBN 0-7277-3455-5.

De Rouck, J., Geeraerts, J., Troch, P., Kortenhaus, A., Pullen, T. & Franco, L. 2005 New results on scale effects for wave overtopping at coastal structures. Proc. Coastlines, Structures & Breakwaters. pp 29–43, ICE London, Thomas Telford, ISBN 0-7277.

De Rouck, J., van der Meer, J. W., Allsop, N. W. H., Franco, L. & Verhaeghe, H. 2002 Wave overtopping at coastal structures: development of a database towards up-graded prediction methods. Proc. 28th Int. Conf. on Coastal Engng (ASCE). Cardiff.

De Rouville, M. A., Bresson, M. M. P. & Petry, P. 1938 État actuel des etudes internationales sur les efforts dus aux lames. Annales des Ports et Chaussées, 108, VII, pp 5–113.

De Waal, J. P. & van der Meer, J. W. 1992 Wave run-up and overtopping on coastal structures. Proc. 23rd Int. Conf. on Coastal Engng. Venice, pp 1758–1771.

De Waal, J. P., Tonjes, P. & van der Meer, J. W. 1996 Overtopping of sea defences. Proc. 25th Int. Conf. on Coastal Engng. Orlando. pp 2216–2229.

Department for Communities and Local Government 2006 Planning Policy Statement 25: Development and Flood Risk. HMSO, London.

Dodd, N. 1998 A numerical model of wave run-up, overtopping and regeneration. Proc ASCE, Jo. Waterway, Port, Coast & Ocean Eng., Vol 124, No 2, pp 73–81, ASCE, New York.

Douglas, S. L. 1984 Irregular wave overtopping rates. Proc. 19th Int. Conf. on Coastal Engng. Houston. pp 316–327.

Douglas, S. L. 1985 Review and comparison of methods for estimating irregular wave overtopping rates. Tech. Report CERC-85, WES, Vicksburg.

EA / Besley, P. 1999 Overtopping of seawalls – design and assessment manual. R & D Technical Report W 178, Environment Agency, Bristol, ISBN 1-85705-069-X.

EAK: 2002 Empfehlungen des Arbeitsausschusses Küstenschutzwerke. Die Küste. H. 65.

FLOODSITE http://www.floodsite.net/html/project_overview.htm.

Forchheimer, P. 1901 Wasserbewegung durch Boden. Z. Ver. Deutsch. Ing., Vol. 45, pp. 1782–1788.

Franco, C. & Franco, L. 1999 Overtopping formulae for caisson breakwaters with non-breaking 3-d waves. Jo. Waterway, Port, Coastal & Ocean Engineering, Vol 125, No 2, ASCE, New York, pp 98–107.

Franco, C., Franco, L., Restano, C. & van der Meer, J. W. 1995 The effect of wave obliquity and short-crestedness on the overtopping rate and volume distribution on caisson breakwaters. Paper 4.9 in MCS Final Report, publn University of Hannover.

Franco, C., van der Meer, J. W., & Franco, L. 1996 Multi-directional wave loads on vertical breakwaters. Proc. 25th Int. Conf. on Coastal Engng. Orlando.

Franco, L., de Gerloni, M. & van der Meer, J. W. 1994 Wave overtopping on vertical and composite breakwaters. Proc. 24th Int. Conf. on Coastal Engng. Kobe pp 1030–1044.

Führböter, A. 1991 Wellenbelastung von Deich- und Deckwerksböschungen. Jahrbuch der Hafenbautechnischen Gesellschaft. Bd. 46. pp 225–228.

Goda, Y. & Morinobu, K. 1998 Breaking wave heights on horizontal bed affected by approach slope. Coastal Engineering Journal, 40, 4, pp 307–326.

Goda, Y. 1967 The fourth order approximation to the pressure of standing waves. Coastal Engineering in Japan, Vol 10, pp 1–11, JSCE, and Tokyo.

Goda, Y. 1971 Expected rate of irregular wave overtopping of seawalls. Coastal Engineering in Japan, Vol 14, pp 45–51, JSCE, Tokyo.

Goda, Y., Kishiya, Y., & Kamiyama, Y. 1975 Laboratory investigation on the overtopping rates of seawalls by irregular waves. Ports and Harbour Research Institute, Vol 14, No. 4, pp 3–44, PHRI, Yokosuka.

Goda, Y. 1985 Random Seas and Design of Maritime Structures. University of Tokyo Press.

Goda, Y. 1969 Reanalysis of Laboratory Data on Wave Transmission Over Breakwaters. Report of the Port and Harbor Institute, Vol. 18. No. 3, PHRI, Yokosuka.

Goda, Y. 2000 Random seas and design of maritime structures (2nd edition). World Scientific Publishing, Singapore, ISBN 981-02-3256-X.

Gonzalez-Escriva, J. A., Garrido, J. M., Medina, J. R., & Geeraerts, J. 2005 Laboratory real storm reproduction using wind. Proc. 29th Int. Conf. on Coastal Engng. Lisbon.

Haehnel, R. B. & Daly, S. F. 2002 Maximum impact force of woody debris on floodplain structures. Technical Report ERDC/CRREL TR-02-2, Cold Regions Research and Engineering Laboratory, ERDC, US Army Corps of Engineers.

Hawkes, P. J. 1999 Mean overtopping rate in swell and bi-modal seas. Technical Note in Proc. ICE, Water, Maritime and Energy, publn. Thomas Telford, London.

Hawkes, P. J., Gouldby, B. P, Tawn, J. A & Owen, M. W. 2000 The joint probability of waves and water levels in coastal defence design. Special Maritime Issue, Jo. Hydraulic Research, IAHR.

Hawkes, P. & Hague, R. 1994 Validation of Joint Probability methods for large waves and high water levels. HR Wallingford Report SR347.

Hedges, T. S. & Reis, M. T. 1998 Random wave overtopping of simple sea walls: a new regression model. Proc. Instn. Civil Engrs. Water, Maritime & Energy, Volume 130, Thomas Telford, London.

Hedges, T. S. & Reis, M. T. 1999 Risk assessment of coastal defences. Proc. 34th MAFF Conference of River & Coastal Engineers, Keele, pp 4.2.1–4.2.13, publn. MAFF, Eastbury House, London.

Hedges, T. S. & Reis, M. T. 2004 Accounting for random wave run-up in overtopping predictions. Maritime Engineering Journal, Proc. ICE, 157(MA3), pp 113–122, Thomas Telford, London.

Hedges, T. S. & Mase, H. 2005 Modified Hunt's equation incorporating wave setup. Journal of Waterways, Port, Coastal and Ocean Engineering. Vol. 130. No. 3 pp. 109–113.

den Heijer, F. 1998 Wave overtopping and forces on vertical water defence structures. WL|Delft Hydraulics. Report H 2014.

Herbert, D. M. 1996 Overtopping of Seawalls, A Comparison between Prototype and Physical Model Data. HR Wallingford Report TR22.

Herbert, D. M., Owen, M. W. & Allsop, N. W. H. 1994 Overtopping of seawalls under random waves. Proc. 24th Int. Conf. on Coastal Engng. Kobe.

Herbert, D. M. 1993 Wave overtopping of vertical walls. Report SR 316, HR Wallingford.

Herbich, J., Sorensen, R. M. & Willenbrock, A. M. 1963 Effect of berm on wave run-up on composite slopes. Journal of the Waterways and harbors devision. WW2. pp 55–72.

Hibberd, S. & Peregrine, D. H. 1979 Surf and run-up on a beach: a uniform bore. Journal of Fluid Mechanics, Vol. 95, and part 2, pp 323–345

Hirt, C. W. & B. D. Nichols 1981 Volume of fluid method for the dynamics of free boundaries. Journal of Comp. Physics, Vol. 39, pp 201–225.

Hu, K., Mingham, C. & Causon, D. M. 2000 Numerical simulation of wave overtopping of coastal structures using the nonlinear shallow water equations. Coastal Engineering 41 (4), pp 433–465.

Hughes, S. A. 2005 Estimating irregular wave run-up on rough, impermeable slopes. US Army Corps of Engineers. ERDC/CHL CHETN-III-70.

Hunt, A. 1959 Design of seawalls and breakwaters. Journal of the Waterways and Harbors Division. pp 123–152.

International Standards Organisation 2003 Actions from waves and currents. ISO/TC98/SC3/WG8.

Juhl, J. & Sloth, P. 1994 Wave overtopping of breakwaters under oblique waves. Proc. 24th Int. Conf. on Coastal Engng. Kobe.

Kajima, R. & Sakakiyama, T. 1994 Review of works using CRIEPI flume and present work. Coastal Dynamics, Barcelona, Spain, pp 614–627.

Kimura, K., Fujiike, T., Kamikubo, K., Abe, R. & Ishimoto, K. 2000 Damage to vehicles on a coastal highway by wave action. Proc. Conf. Coastal Structures, Santander, publn. A.A. Balkema, Rotterdam.

Kobayashi, N. & Raichle, A. 1994 Irregular wave overtopping revetments in surf zones. Jo. Waterway, Harbour and Coastal Eng. Div., Proc. ASCE, Vol. 120, No 1, pp 56–73, publn. ASCE, New York.

Kobayashi, N. & Wurjanto, A. 1989 Wave overtopping on coastal structures. Jo. Waterway, Harbour and Coastal Eng. Div., Proc. ASCE, Vol. 115, No 2, pp 235–251, publn. ASCE, New York.

Kobayashi, N., Otta, A. K. & Roy, I. 1987 Wave reflection and run-up on rough slopes. J. of Waterway, Port, Coastal and Ocean Engineering, ASCE, Vol. 113, No. 3, pp 282–298.

Kortenhaus, A., Oumeraci, H., Geeraerts, J., De Rouck, J., Medina, J. R., & Gonzalez-Escriva, J. A. 2004a Laboratory effects and other uncertainties in wave overtopping measurements. Proc. 29th Int. Conf. on Coastal Engng. Lisbon.

Kortenhaus, A., van der Meer, J. W., Burcharth, H., Geeraerts, J., van Gent, M. & Pullen, T. 2004b Final report on scale effects. CLASH WP7-report, LWI, and Germany.

Kortenhaus, A. Haupt, R. & Oumeraci, H. 2001 Design aspects of vertical walls with steep foreland slopes. Proc Breakwaters, coastal structures and coastlines, London (ICE), pp 221–232 (ISBN 0-7277-3042-8)

Kortenhaus, A., Pearson, J., Bruce, T., Allsop, N. W. H. & van der Meer, J. W. 2003 Influence of parapets and recurves on wave overtopping and wave loading of complex vertical walls. Proc ,'Coastal Structures 2003", pp 369–381, ASCE, Reston, Virginia, ISBN0-7844-0733-9

Mayer, R. H. & Kriebel, D. L. 1994 Wave Run-up on composite slope and concave beaches. Proc. 24th Int. Conf. on Coastal Engng. Kobe. pp 2325–2339.

McConnell K. J. 1998 Revetment systems against wave attack: a design manual. ISBN 0-7277-2706-0, Thomas Telford, London.

Mendez-Lorenzo, A. B., van der Meer, J. W. & Hawkes, P.J. 2000 Effects of bi-modal waves on overtopping: application of UK and Dutch prediction methods. Proc. 27th Int. Conf. on Coastal Engng. Sydney.

Monaghan, J. J. 1994 Simulating free surface flows with SPH. J. of Comp. Physics, 110: 499–406.

Moriya, Y. & Mizuguchi, M. 1996 Wave overtopping rate and reflection coefficient for obliquely incident waves. Proc. 25th Int. Conf. on Coastal Engng. Orlando. pp 2598–2611.

Napp, N., Pearson, J., Bruce, T. & Allsop, W. 2002 Overtopping of seawalls under oblique wave attack and at corners. Proc. 28th Int. Conf. on Coastal Engng. Cardiff. pp 2178–2190.

Napp, N., Bruce, T., Pearson, J. & Allsop, N. W. H. 2004 Violent overtopping of vertical seawalls under oblique wave conditions. Proc. 29th Int. Conf. on Coastal Engng. Lisbon. pp 4482–4493

Oumeraci, H. 1999 Strengths an limitations of physical modelling in coastal engineering-synergy effects with numerical modelling and field measurements. Proceedings Hydralab Workshop on Experimental Research and Synergy Effects with Mathematical Models, Evers, K.-U., Grüne, J., Van Os, A. (eds), Hannover, Germany, pp 7–38.

Oumeraci, H., Kortenhaus, A., Allsop, N. W. H., de Groot, M. B., Crouch, R. S., Vrijling, J. K. & Voortman, H. G. 2001 Probabilistic design tools for vertical breakwaters. A A Balkema, Rotterdam, ISBN 90-580-248-8.

Oumeraci, H., Möller, J., Kübler, S., Zimmermann, C., Daemrich, K. F. & Ohle, N. 2002 Einfluss von Bermen und Knicken im Deichquerschnitt auf den schrägen Wellenauflauf. LWI-Report No. 880.

Oumeraci, H., Schüttrumpf, H. & Bleck, M. 1999 Untersuchungen zur Ermittlung der mittleren Wellenüberlaufrate ohne Freibord bei Stromdeichen. LWI-Report No. 842.

Oumeraci, H., Schüttrumpf, H., Möller, J. & Kudella, M. 2001 Loading of the inner slope of seadikes by wave overtopping – results from large scale model tests. LWI-Report No. 858.

Owen, M. W. & Steele, A. A. J. 1991 Effectiveness of recurved wave return walls. HR Wallingford, Report SR 261.

Owen, M. W. 1980 Design of seawalls allowing for wave overtopping. HR Wallingford, Report EX 924.

Owen, M. W. 1982 Overtopping of Sea Defences. International Conference on the Hydraulic Modelling of Civil Engineering Structures, Coventry.

Owen, M. W. 1982 The hydraulic design of sea-wall profiles. Proc. ICE Conf. on Shoreline Protection. pp 185–192, publn Thomas Telford, London

Owen, M. W., Hawkes, P. J., Tawn, J. A & Bortot, P. 1997 The joint probability of waves and water levels:a rigorous but practical new approach. Proc. MAFF Keele Conf. of River & Coastal Engineers, MAFF, London.

Pearson, J., Bruce, T., Allsop, N. W. H. & Gironella, X. 2002 Violent wave overtopping – measurements at large and small scale. Proc. 28th Int. Conf. on Coastal Engng. Cardiff. pp 2227–2238.

Pearson, J., Bruce, T., Allsop, N. W. H., Kortenhaus, A. & van der Meer, J. W. 2004 Effectiveness of recurve wave walls in reducing wave overtopping on seawalls and breakwaters. Proc. 29th Int. Conf. on Coastal Engng. Lisbon. pp 4404–4416.

Pedersen, J. & Burcharth, H. F. 1992 Wave forces on crown walls. Proc. 23rd Int. Conf. on Coastal Engng. Venice.

Pedersen, J. 1996 Wave forces and overtopping on crown walls of rubble mound breakwaters – an experimental study. Series paper 12, ISSN 0909-4296, Hydraulics and Coastal Eng. Lab., Aalborg University, Denmark

Petit, H. A. H. & van den Bosch, P. 1992 SKYLLA: Wave motion in and on coastal structures; Numerical analysis of program modifications. Delft Hydraulics Report H1351.

Pozueta, B., van Gent, M. R. A, van den Boogaard, H. & Medina, J. R. 2004 Neural network modelling of wave overtopping at coastal structures. Proc. 29th Int. Conf. on Coastal Engng. Lisbon. pp 4275–4287.

Pullen, T., Allsop, N. W. H., Bruce, T. & Geeraerts, J. 2003 Violent wave overtopping: CLASH field measurements at Samphire Hoe. Proc Coastal Structures 2003 conf, Portland, pp 469–480, ASCE, New York.

Pullen, T., Allsop, N. W. H., Pearson, J. & Bruce, T. 2004 Violent wave overtopping discharges and the safe use of seawalls. Proc. Defra Flood & Coastal Management Conf., York, publn. Flood Management Division, Department for Environment Food and Rural Affairs, London.

Pullen, T., Allsop, N. W. H., Bruce, T., Pearson, J. & Geeraerts, J. 2004 Violent wave overtopping at Samphire Hoe: field and laboratory measurements. Proc. 29th Int. Conf. on Coastal Engng. Lisbon. pp 4379–4390.

Regeling, H. J., van der Meer, J. W., 't Hart, R. & Bruce, T. 2005 Overtopping on rock berm with smooth upper slope. Proc. Second International Coastal Symposium, Höfn, Iceland.

Richardson, S. R., Ingram, D. I., Mingham, C. G. & Causon, D. M. 2001 On the validity of the shallow water equations for violent overtopping. Proc. Conf. Waves 2001, pp 1112–1125, publn. ASCE, New York.

Richardson, S., Pullen, T. & Clarke, S. 2002 Jet velocities of overtopping waves on sloping structures: measurements and computation. Proc. 28th Int. Conf. on Coastal Engng. Cardiff.

Roos, A. and Battjes, J.A. 1976 Characteristics of Flow in Run-up of periodic waves. Proc. 15th Int. Conf. on Coastal Engng. Honolulu. pp 781–795.

RWS. 2001 Guidance on hydraulic boundary conditions for the safety assessment of Dutch water defences. Hydraulische Randvoorwaarden.

Sabeur, Z., Allsop, N. W. H., Beale, R. G. & Dennis, J. M. 1996 Wave dynamics at coastal structures: development of numerical models of wave kinematics. Proc. 25th Int. Conf. on Coastal Engng. Orlando. pp 389–402.

Saville, T. 1958 Wave Run-up on Composite Slopes. Proc. 6th Int. Conf. on Coastal Engng. Gainesville. pp 691–699.

Schiach, J., Mingham, C. G., Ingram, D. M., Causon, D. M, Bruce, T, Pearson, J. P. & Allsop, N. W. H. 2004 Extended shallow water prediction of overtopping. Proc. 29th Int. Conf. on Coastal Engng. Lisbon. pp 4443–4455.

Schulz, K-P. 1992 Massstabseffekte beim Wellenauflauf auf glatten und rauhen Böschungen. Mitteilungen Leichtweiss-Institut für Wasserbau der Technischen Universität Braunschweig, Heft 120, Braunschweig, Germany.

Schüttrumpf, H. & van Gent, M. R. A. 2003 Wave overtopping at seadikes. ASCE, Proc. Coastal Structures 2003, Portland, USA, pp 431–443.

Schüttrumpf, H. 2003 Wave overtopping flow on seadikes – Experimental and theoretical investigations. PIANC Bulletin 149.

Schüttrumpf, H., Bergmann, H., Dette, H. H. 1994 The concept of residence time for the description of wave run-up, wave set-up and wave run-down. Proc. 24th Int. Conf. on Coastal Engng. Kobe. pp 553–564.

Schuttrumpf, H. & Oumeraci, H. 1999 Wave overtopping at seadykes. Proc. HYDRALAB workshop, Hannover, pp 327–334, ISBN 3-00-004942-8, publn. Forschungszentrum Küste, Hannover.

Schüttrumpf, H. & Oumeraci, H. 2005 Layer Thicknesses and Velocities of Wave Overtopping at Seadykes. Journal of Coastal Engineering. No. 52. pp 473–495.

Schüttrumpf, H. 2001 Wellenüberlaufströmung bei Seedeichen – Experimentelle und Theoretische Untersuchungen. PhD-Thesis.

Schüttrumpf, H., Barthel, V., Ohle, N., Möller, J. & Daemrich, K. F. 2003 Run-up of oblique waves on sloped structures. COPEDEC VI Conference. Colombo. Sri Lanka.

Schüttrumpf, H., Möller, J. & Oumeraci, H. 2002 Overtopping Flow Parameters on the inner slope of seadikes. Proc. 28th Int. Conf. on Coastal Engng. Cardiff. pp 2116–2128.

Schüttrumpf, H., Möller, J., Oumeraci, H., Grüne, J, & Weissmann, R. 2001b Effects of Natural Sea States on Wave Overtopping of Seadikes. Proceedings Waves 2001 Conference. San Francisco. pp 1565–1574.

Schüttrumpf, H., Oumeraci, H., Thorenz, F. & Möller, J. 2001a Reconstruction and Rehabilitation of a historical Seawall at Norderney. Proceedings Coastlines, Structures and Breakwaters Conference. London. Thomas-Telford-Verlag. pp 257–268.

Schüttrumpf, H. & Oumeraci, H. 2005 Scale and Model effects in crest level design. Proc. 2nd Coastal Symposium. Höfn. Iceland.

Schüttrumpf, H., Kortenhaus, A., Petes, K. & Fröhle, P. 2006 Expert Judgement of Uncertainties in Coastal Structure design. Proc. 30th Int. Conf. on Coastal Engng. San Diego.

Simm, J. D., Brampton, A. H., Beech, N. W. & Brooke, J. S. 1996 Beach management manual. Report 153, ISBN: 0-86017 438 7, CIRIA, London.

SKYLLA Delft Hydraulics VOF Model. www.wldelft.nl/soft/SKYLLA/index.html

Smid, R. 2001 Untersuchungen zur Ermittlung der mittleren Wellenüberlaufrate an einer senkrechten Wand und einer 1:1,5 geneigten Böschung für Versuche mit und ohne Freibord. Student study at Leichtweiss-Institute for Hydraulics. Braunschweig (in German)

Smith, G. M., Seijffert, J. W. W. & van der Meer, J. W. 1994 Erosion and Overtopping of a Grass Dike, Large Scale Model Tests. Proc. 24th Int. Conf. on Coastal Engng. Kobe.

Steendam, G. J., van der Meer, J. W., Verhaeghe, H., Besley, P., Franco, L. & van Gent, M.R.A. 2004 The international database on wave overtopping. Proc. 29th Int. Conf. on Coastal Engng. Lisbon. pp 4301–4313.

Stewart, T., Newberry, S., Latham, J-P & Simm, J. D. 2003 Packing and voids for rock armour in breakwaters. Report SR 621, HR Wallingford.

Stewart, T., Newberry, S., Simm, J. & Latham, J-P. 2002 Hydraulic performance of tightly packed rock armour - results from random wave model tests of armour stability and overtopping. Proc. 28th Int. Conf. on Coastal Engng. Cardiff. pp 1449–1461.

Stickland, I. W. & Haken, I. 1986 Seawalls, Survey of Performance and Design Practice. Tech Note No. 125, ISBN 0-86017-266-X, Construction Industry Research and Information Association (CIRIA) London.

Sutherland, J. & Gouldby, B. 2003 Vulnerability of coastal defences to climate change. Proc. ICE, Water & Maritime Engineering Vol. 156, Issue WM2, pp 137–145 (Thomas Telford, London).

Szmytkiewicz, M., Zeidler, R. & Pilarczyk, K. 1994 Irregular Wave Run-up on Composite Rough Slopes. Coastal Dynamics, pp 599–613.

Tautenhain, E. 1981 Der Wellenüberlauf an Seedeichen unter Berücksichtigung des Wellenauflaufs. Mitt. des Franzius-Instituts. No. 53. pp 1–245

TAW: 1997 Technical Report – Erosion Resistance of grassland as dike covering. Technical Advisory Committee for Flood Defence in the Netherlands (TAW). Delft. 1997

TAW: 2002 Technical Report – Wave run-up and wave overtopping at dikes. Technical Advisory Committee for Flood Defence in the Netherlands (TAW). Delft. 2002

Thomas, R. S. & Hall, B. 1992 Seawall design. ISBN 0-7506-1053-0, CIRIA / Butterworth-Heinemann, Oxford.

Van der Meer, J. W., Snijders, W. & Regeling, H. J. 2006 The wave overtopping simulator. Proc. 30th Int. Conf. on Coastal Engng. San Diego.

Van der Meer, J. W., van Gent, M. R. A., Pozueta, B., Verhaeghe, H., Steendam, G-J., Medina, J. R. 2005 Applications of a neural network to predict wave overtopping at coastal structures. ICE, Coastlines, Structures and Breakwaters, London, pp 259–268.

Van der Meer, J. W. 1998 Wave run-up and overtopping. Chapter 8 in: "Seawalls, dikes and revetments." Edited by K.W. Pilarczyk. Balkema, Rotterdam.

Van der Meer, J. W., Briganti, R., Zanuttigh, B. & Wang B. 2005 Wave transmission and reflection at low-crested structures: Design formulae, oblique wave attack and spectral change. Special Issue of Journal of Coastal Engineering, Elsevier, 52, pp 915–929.

Van der Meer, J. W., Briganti, R., Wang, B. & Zanuttigh, B. 2004 Wave transmission at low-crested structures, including oblique wave attack. Proc. 29th Int. Conf. on Coastal Engng. Lisbon. pp 4152–4164.

Van der Meer, J. W., Wang, B., Wolters, A., Zanuttigh, B. & Kramer, M. 2003 Oblique wave transmission over low-crested structures. ASCE, Proc. Coastal Structures, Portland, Oregon, pp 567–579.

Van der Meer, J. W. & de Waal, J. P. 1992 Summary of wave overtopping at dikes. Technical Note, Delft Hydraulics.

Van der Meer, J. W., Tönjes, P. & de Waal, J. P. 1998 A code for dike height design and examination. Proc. Conf. Coastlines, Structures & Breakwaters 1998, Institution of Civil Engineers, pp 5–21, publn. Thomas Telford, London.

Van der Meer, J. W. & Janssen, J. P. F. M. 1995 Wave run-up and wave overtopping at dikes. Chapter 1 in ‚'Wave Forces on Inclined and Vertical Wall Structures", pp 1–26, ed. Kobayashi, N. & Demirbilek, Z., ASCE, New York, ISBN 0-7844-0080-6.

Van der Meer, J. W. & de Waal, J. P. 1990 Influence of oblique wave attack and short-crested waves on wave run-up and wave overtopping. WL|Delft Hydraulics. H 638.

Van der Meer, J. W. and Janssen, P. F. M. 1995 Wave Run-up and Wave Overtopping at Dikes. ASCE book on „Wave Forces on inclined and vertical wall structures", Ed. Z. Demirbilek.

Van der Meer, J. W. & Klein Breteler, M. 1990 Measurement and computation of wave induced velocities on a smooth slope. Proc. 22nd Int. Conf. on Coastal Engng. Delft. pp 191–204.

Van der Meer, J. W. 1988 Rock slopes and gravel beaches under wave attack. PhD-Thesis Delft University of Technology. Also Delft Hydraulics Publications. No. 396.

Van Gent, M. R. A. 1999 Physical Model Investigations on Coastal Structures with shallow foreshores – 2D model tests with single and double peaked wave energy spectra. Delft Hydraulics. Report H. 3608.

Van Gent, M. R. A. 1995 Wave interaction with permeable coastal structures. PHD-thesis. Delft University Press.

Van Gent, M. R. A. 2002 Wave overtopping events at dikes. Proc. 28th Int. Conf. on Coastal Engng. Cardiff. pp 2203–2215.

Van Gent, M. R. A. 2000 Wave run-up on dikes with berms. WL|Delft Hydraulics H3205.

Van Gent, M. R. A. 2001 Wave runup on dikes with shallow foreshores. Journal of Waterway, Port, coastal and Ocean Engineering. Vol. 127. No. 5. pp 254–262.

Van Gent, M. R. A. 1994 The modelling of wave action on and in coastal structures. Coastal Engng. 22, pp 311–339.

Van Gent, M. R. A., Petit, H. A. H. & van den Bosch, P. 1994b SKYLLA: Wave motion in an on coastal structures; Implementation and verification of flow on and in permeable structures. Delft Hydraulics Report H1780

Van Gent, M. R. A. 1991 Formulae to describe porous flow. Communications on Hydraulic and Geotechnical Engineering, ISSN 0169-6548 No. 92-2, Delft University of Technology and MAST-G6S report.

Van Gent, M. R. A. 1993 Manual on ODIFLOCS (version 2.0). Delft University of Technology.

Van Gent, M. R. A. 1995 Wave interaction with permeable coastal structures. Communications on Hydraulic and Geotechnical Engineering, Report No. 95-5, Faculty of Civil Engineering, Delft University of Technology (PhD Thesis).

Van Gent, M. R. A., de Waal, J.P., Petit, H.A.H. & van den Bosch, P. 1994a SKYLLA: Wave motion in an on coastal structures; Verification of wave kinematics of waves breaking on an offshore bar. Delft Hydraulics Report H1780.

Verhaeghe, H., van der Meer, J.W., Steendam, G-J., Besley, P., Franco, L. &. van Gent, M. R. A 2003 Wave overtopping database as the starting point for a neural network prediction method. ASCE, Proc. Coastal Structures 2003, Portland, Oregon, pp 418–430.

Ward, D. L., Wibner, C. G., Zhang, J., & Edge, B. 1995 Wind effects on runup and overtopping. Proc. 24th Int. Conf. on Coastal Engng. Kobe.

Ward, D., Zhang, J., Wibner, C. & Cinotto, C. 1996 Wind effects on Runup and Overtopping of Coastal Structures. Proc. 25th Int. Conf. on Coastal Engng. Orlando. pp 2206–2215.

Wassing, F. 1957 Model investigations on wave run-up carried out in the Netherlands during the past twenty years. Proc. 6th Int. Conf. on Coastal Engng. Gainesville. pp 700–714.

Weston, B. P., Borthwick, A. G. L., Taylor, P. H., Hunt, A. C. & Stansby, P. K. 2005 Performance of a Hybrid Boussinesq Model on Wave Runup and Overtopping Predictions for Coastal Structures. Paper to ICE conf. Coastal Structures & Breakwaters 2005, London.

Wolters, G., Müller, G., Bruce, T. & Obhrai, C. 2005 Large scale experiments on wave downfall pressures on vertical and steep coastal structures. Proc. ICE, Maritime Engineering, 158, pp 137–145.

A Structure of the EurOtop calculation tool

To complement the EurOtop manual, a website has been designed to simplify the empirical formula by giving the user a choice of standard structures to calculate overtopping rates. The EurOtop calculation tool can be found at http://www.overtopping-manual.com.

It is intended for with basic structures only for more complex situations please use the software PC Overtop or use the neural network.

Calculation tool home page

The introduction page contains a list of the most popular structures and the methods available to calculate overtopping discharge. PC Overtopping and the neural network method instructions are describes elsewhere in the manual.

To calculate overtopping discharge click the empirical method link next to the desired structure or alternatively select the Empirical Methods tab for a full list of structures.

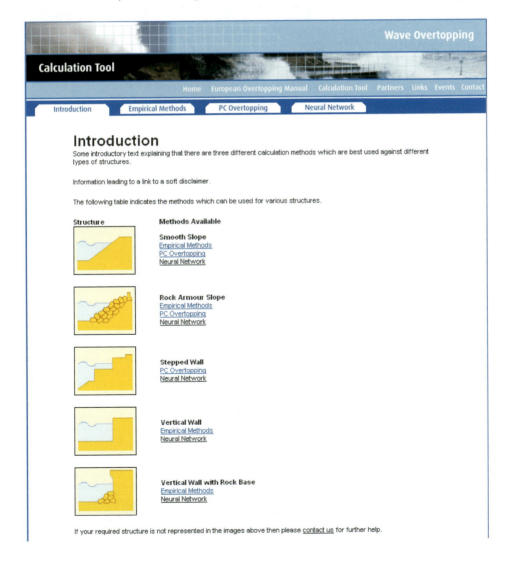

Empirical Methods Page

The empirical method page contains most structure types currently available. These are designed to follow the guidelines set out in Chapters 5–7 of the manual. If no basic type exists for your desired structure then use one of the other methods by selecting the introduction tab (Refer to Chapter 4).

To calculate overtopping rates click the relevant structure type.

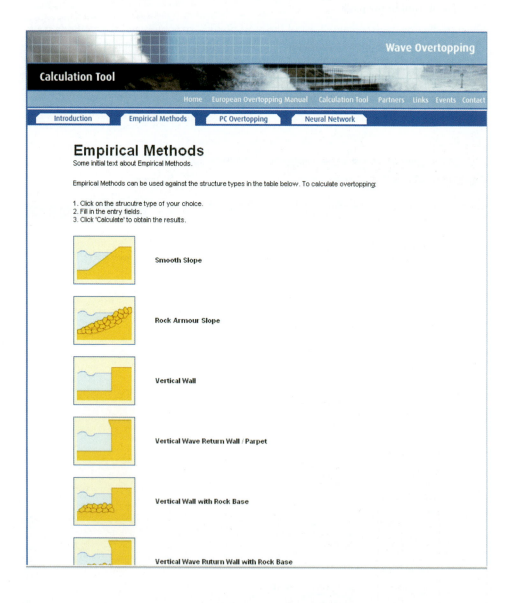

Overtopping calculation

Once a structure type has been chosen the calculation page will be displayed.

1. Input

Each structure type will have different input variables and all require a wave period, freeboard and wave height. The wave period, **T**, can be input either as a mean (T_m), peak (T_p) or $T_{m-1,0}$. This spectral period $T_{m-1,0}$ gives more weight to the longer wave periods in the spectrum and is therefore well suited for all kind of wave spectra including bi-modal and multi-peak wave spectra.

The freeboard (**Rc**) is simply the height of the crest of the wall above still water level. A wave height at the toe of the structure (H_{m0}) is also needed for most calculations. Sloped structures also contain a reduction factor (**γ**). A range of materials are listed along with armour based slopes. Please refer to the manual for guidance if no material type exists for your structure.

All variables must be entered before an overtopping rate can be calculated, for help on any variable please refer to the manual.

An example Input screen for a vertical wall structure is shown below.

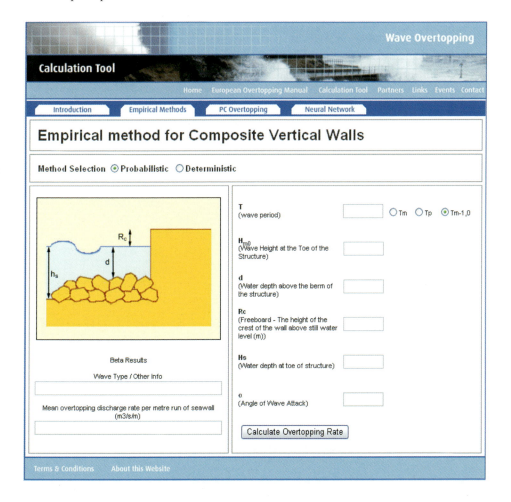

2. Output

There are two outputs from the calculations, an overtopping rate and a structure specific comment about the calculation method.

The overtopping rate is listed as metres / second mean overtopping discharge per meter structure width [$m^3/s/m$].

The comment box will list any observations or errors from the formulae, these can range from wave breaking type (sloped structures) to impulsive waves (vertical structures).

For interpretation of the results please consult the Eurotop manual.